DO GENTLEMEN REALLY PREFER BLONDES?

DO GENTLEMEN REALLY PREFER BLONDES?

Why He Fancies You And Why He Doesn't

JENA PINCOTT

BANTAM PRESS

LONDON • TORONTO • SYDNEY • AUCKLAND • JOHANNESBURG

A CIP catalogue record for this book is available from the British Library.

ISBN 9780593060100

Addresses for Random House Group Ltd companies outside the UK can be found at: www.randomhouse.co.uk
The Random House Group Ltd Reg. No. 954009

The Random House Group Limited supports The Forest Stewardship Council (FSC), the leading international forest-certification organization. All our titles that are printed on Greenpeace-approved FSC-certified paper carry the FSC logo. Our paper procurement policy can be found at www.rbooks.co.uk/environment

Typeset in 12.5/16pt Fournier MT
Printed in the UK by CPI Mackays, Chatham, ME5 8TD

2 4 6 8 10 9 7 5 3 1

CONTENTS

INTRODUCTION

One fall evening, as the air turned bitterly cold and the threat of lonely holidays loomed, my friend Rita went speed-dating. Rita is beautiful and vivacious, and she was up to the challenge of meeting more men in an hour than most women meet in a year. At first she considered one of the special-interest sessions: "Theater Lovers," "Fitness and Health Lovers," or the one called "Doggie Style," which turned out to be for canine owners. "I'm going to find my husband," she said when I raised a skeptical eyebrow. She settled on a session for thirtysomethings, and went by herself because her friends refused to go with her. The next day Rita was in an expansive, exuberant mood. She lay on her back with her hand behind her head, giggling and gazing at the ceiling. She'd met twenty men, she said, in a sort of fevered rush. Three minutes was all she'd had to form an impression of each "date" before the bell rang and she moved on to the next. One guy outshined the others. They had a *fabulous* connection. "Your future husband?" I asked, impressed.

Rita says she's looking for a man who's loyal, responsible, educated, spiritual, and ambitious and who wants to be a father. That's the rational Rita speaking. But in the heat of the moment she forgets her intentions. Rita rolled her eyes at the earnest duds who dressed up to speed-date. Instead, her marvelous man, the only one to whom she said yes, turned out to be a sweet-talking brute who was living on his friend's couch. For three minutes he gazed at her with burning eyes and asked questions like "Sweetheart, why don't you model?" Maybe he wasn't marriage material, but she was smitten.

After Rita's experience, it didn't shock me to learn that in every study on speed-dating, men and women's self-reported mate preferences are unrelated to the characteristics of the people they actually pick. We often rely on instinct or urge more than reason. In fact, half of all female speed-daters say they know whether they're going to say yes to a guy within the first *three seconds* of meeting him. Men are also startlingly efficient, and both sexes care a lot about looks. By the time the bell rings, all the participants have made up their minds.

So, what happens in those three seconds or three minutes? What part of Rita—or you—decides what's sexy? Not the rational brain. When it comes to attraction, consciousness slips down a gear. The instincts go into overdrive. The senses take over. Unconsciously, you're taking in the timbre of your date's voice, the sturdiness of his shoulders, the thickness of his brow and jaw, the good humor in his gaze. *Looks right, sounds right, smells right, acts right.* You might feel a slow, burning blush. You find yourself leaning in his direction. It's as if your body's doing the deciding—your eyes, ears, nose, hormones, or something deep in the back of your brain.

All the time—but especially in your love life—you're making decisions beyond your conscious awareness, and people respond to you in ways and for reasons unconscious to them. There might be days when you find yourself acting a little more flirtatious. This morning, on a whim, you might have decided to wear a sexier outfit than usual. Your skin is softer, your features more symmetrical. Men seem to be drawn to you. You find yourself opening up like a flower when talking to cocky, domineering guys, even though they're not normally your type. What's going on? (See page 122.)

It turns out that there are many deep and subtle influences that draw you to certain people, and they to you. Or not! Take body odor as an example. Why is it that you love the smell of some men's sweat but not others? A man's natural odor is a make-it-or-break-it factor for many women. In fact, its surprising sway was the inspiration behind this book. I once dated a guy whose smell I hated, even though he showered, and it was a major reason why I couldn't take the relationship further. Later on, I met a man whose smell I love—and I married him (for that and his other amazing qualities). When I found out there's a biological basis to my olfactory pickiness, I was intrigued. (See page 30.)

There's real science behind a lot of odd, under-the-radar things that happen in your love life, such as why you climax more often with some lovers than with others, why sex makes you feel sated, and why cuddling with a guy makes you a little more attached and trusting, even when you don't want to be. There are reasons why men think you're into them when you're not; why people seem more attractive when you're excited or when you gaze into their eyes; why going on the Pill could change your taste in men; why your sex drive may pick up in autumn; and why you get so crazy when you fall in love.

Of course, men also are egged on by urges and instincts. There are reasons why they have a different reaction to pornography than women do, why they get muddleheaded at the sight of beautiful babes, and why they get so amorous after you spend time away from them. You might also wonder why so many guys are enchanted by voluptuous breasts, hourglass figures, and long legs. And what's the big deal with blond hair?

Do Gentlemen Really Prefer Blondes? explores the hidden side of love, sex, and attraction. The questions in this book—nearly one hundred—were driven by my somewhat insatiable curiosity about science, sex appeal, and the subconscious. *What goes on that no one talks about because we hardly know it happens?* For answers I searched hundreds of peer-reviewed studies in diverse disciplines: biology, evolutionary psychology, anthropology, neuroscience, endocrinology, and others. I was fascinated.

As a writer with a science background, yet a nonspecialist, I cast the net far and wide, finding topics ranging from body language to bisexuality, hormones to pheromones, and "sexy genes" to "mate value" models. Drawing on these studies, with insights from interviews with many of the researchers, this book showcases all the research that caught my eye about attraction and its aftermath, love and sex (not necessarily in that order). Some of the findings featured here are mainstream, while others are recent and more controversial. Although no one study solves the mysteries of love and attraction, each is something of a clue. Together, they provide a "big-picture" perspective. (Scientists don't really believe that our love lives can be reduced to science, but that we can use it to understand ourselves better.)

One theme in this romp through the research is that everyone has an unconscious preference for certain traits, and much of what we desire is rooted in deep evolutionary biases. We *evolved* this way. Studying ancestral conditions and the mating behaviors of other animals, evolutionary biologists have an interesting take: whether or not you actually *want* kids, you have "parental investment" instincts that impact your sex life. It all boils down to the basic biological truth that in one year's time, a woman could sleep with a googol of men but only have one full-term pregnancy, whereas a man could sleep with a googol of women and have googols of babies. To maximize their reproductive success, men are attracted to cues of fertility—youth and beauty—especially in short-term relationships. For women, it's more complicated. Women have more at stake in the event of a pregnancy, so we're choosier about our sex partners. Over the ages we developed biases for guys with signs of good genes (masculinity and social or physical dominance) and signs that they would be good dads (nurturers and providers), although we often make trade-offs depending upon our circumstances. While it's a given that culture and personal experience affect the decisions we make in our love (and sex) lives, the hidden forces of urges and instincts influence us unexpectedly.

There are so many questions. Why do big-mouthed, broad-shouldered guys attract my friend Rita, and what about her attracts them? How do men get into your pants, or your heart, and you into theirs? And why, when you're with the right person, are love and sex so mind-blowing? In writing this book I was delighted to learn that so many researchers in so many fields are exploring topics relevant to our love lives—from how we look to how we smell, and from why we make love to how we stay in love. (This is a browser's book, so flip through the questions and let your interests guide you.) These discoveries provide helpful insights into human nature. Better yet, they're a lot of fun. As the Nobel Prize–winning physicist Richard Feynman put it, "Science is a lot like sex. Sometimes something useful comes of it, but that's not the only reason we're doing it."

PART I
BODIES

CHAPTER 1 *Face First*

Who ever loved, that loved not at first sight?

—Christopher Marlowe, from "Hero and Leander"

Why do people seem more attractive when you are gazing into their eyes?

Many years ago the behavioral psychologist Arthur Aron put opposite-sex college students in a room and asked them to reveal intimate details of their lives: their most embarrassing moments, what they'd do if their parents died, and so on. Then he paired them up, man and woman, and told them to lock eyeballs for four minutes. No talking, no smiling—just gazing. Deep gazing, like lovers. Later, Aron quizzed the students on how they felt about their partners. Deeply attracted, most said. So deeply that a couple that were strangers on the day of the experiment allegedly got married six months later. You might open the heart by sharing intimacies, but, evidently, you reach it through the eyes.

Looking directly into a lover's eyes is like looking into fire. As Nietzsche put it, "If you gaze for long into an abyss, the abyss gazes also into you." Thanks to a shot of adrenaline, your palms sweat, your breathing gets shallow, your skin feels hot, and your pupils dilate. Your amygdala, the center of the brain that processes emotion, blazes with activity. At the same time you produce dopamine, a "feel-good" neurotransmitter that is associated with passion and addiction, and oxytocin, a hormone related to bonding. So intense is the mutual gaze that there's only one way to amp it up: deeply penetrate your partner's eyes during slow rhythmic sex, as prescribed in the *Kama Sutra* (not recommended with a stranger).

The most fascinating theory about eye gaze is that just the act of doing it can enhance, or even initiate, a feeling of love. Most of the time we think that our faces reflect what's going on inside our heads, but, for at least some people, the expression on their face becomes a genuine feeling. Psychologists call this facial feedback, and Darwin was among its first believers.

The facial feedback hypothesis was borne out in experiments at Clark University and the University of Alaska. At Clark, more than seventy opposite-sex strangers, under the pretext of an ESP study, silently gazed into each other's eyes for two minutes. Participants who were previously assessed and known to respond emotionally to their own facial expressions reported a significant increase in passionate love for the strangers in whose eyes they had gazed. (The gaze must be mutual and nonthreatening.) At the University of Alaska, eye-gazers who scored high on a standard psychological test known as the Romantic Beliefs Scale had the same experience. Men and women who felt strongly about concepts such as a "one and only love," "love at first sight," and "love will find a way" felt a significant surge of romantic love after locking eyes.

According to the facial feedback hypothesis, you might feel tenderhearted toward people after gazing into their eyes because you're acting like a person in love. If anyone saw you softly gazing at someone, they'd think you were in love. You're conscious of the facial muscles that create the expression on your face that everyone sees, and you internalize it. Your neural reward circuits fire up, and you *feel* how you *act*. It's no wonder that professional actors whose characters are in love often fall in love with each other on the set. Of course, causal facial feedback works only if you're aware of and respond to your personal bodily cues. Not everyone does; you may need to be primed by a previous emotional connection or have strong romantic beliefs.

Assuming you are a romantic, gazing into another person's eyes makes that person appear more attractive to you, and just might help you fall in love. But is it *real* love? That remains to be seen.

Why do men prefer big pupils?

The film *Kinsey*, a biopic about the provocative twentieth-century sex researcher Alfred Kinsey, reenacts a moment from one of his famous lectures on the human body. Kinsey turns to a prudish young woman and asks, "What organ in the human body can expand its size a hundred times?" She blushes crimson. "I'm certain I wouldn't know," she replies. Kinsey raises his eyebrows. "I'm talking about the iris of your eye," he says chidingly, and remarks that the young lady could be disappointed if she persists in her way of thinking.

It turns out the iris and the pupil are erotic in their own right. The iris is the colored part of the eye surrounding the pupil, and the pupil is the black "bull's-eye" of the eyeball. The muscles of

Try the Gaze

A close friend once invited me to attend a "transformational" seminar with her. Among other touchy-feely activities, we were told to gaze for two minutes into the eyes of strangers. It was incredibly difficult to do at first, as odd and as intimate as undressing in public. But it worked—more than a decade later I still remember the unexpected bubble of affection that rose in me when I gazed into those strangers' eyes.

Why not try a moment of extended eye contact? Do it with someone you already love, or, if you're daring, try it on a date. In conversation, catch the person's eye and hold it for a beat or two longer than you would otherwise. Look away and look back as you talk, stretching your moments of eye contact longer and longer. It takes courage to hold another's gaze—but if you're the type to believe that the eyes lead to the heart, here's your chance.

the iris expand and contract the pupil from less than 1 mm to nearly 10 millimeters in diameter. The pupil is arguably the face's most blatant and bewitching feature. A wide-open pupil intensifies a person's gaze. Those big black holes suck you in, like it or not.

Everyone knows that pupils dilate in the dark, but there's more to it than meets the eye. In the 1960s, the psychologist

Eckhard Hess discovered that pupils also dilate when people are aroused or emotionally charged. Women's pupils dilated when they saw images of children or male nudes, and men's pupils dilated when they saw female nudes. It's an involuntary reflex of the sympathetic nervous system.

Pupil size is also detected unconsciously. When Hess asked men to judge two pictures of a woman that were identical in every way but the woman's pupil size, the guys overwhelmingly preferred the version with larger pupils. Forced to explain why they thought the woman was more attractive in that picture, the men shrugged and said she just seemed prettier and more feminine. No one consciously noticed the difference in her pupils.

Evolutionarily speaking, men prefer big, gaping pupils because they're a sign of arousal and receptivity. If your pupils are dilated when you're talking to a guy (and you're not drunk or drugged), it's a sign that you're attracted to him. Your pupils dilate widest around ovulation, the fertile phase of the menstrual cycle, and when you're fairly young. As you grow older, your pupils can't dilate as much as they did in childhood and young adulthood. Big pupils are cues of youth, fertility, and receptivity—in the subconscious male mind, a sight to behold.

Women, meanwhile, are less enthusiastic about men with big pupils. A study at York University in Canada found that gals prefer guys with medium-sized pupils. While men regard pupil dilation as a promising sign of arousal, women are often suspicious of it. A wound-up, wild-eyed guy might force you to have sex, or he might be madly overpossessive or somehow out of control. (The few women who preferred men with big pupils tended to also prefer "bad boys.") If a man's pupils are too big, a woman's might contract.

What makes a face good-looking?

While you're walking down the street, you pass a person so drop dead gorgeous that everyone—male, female, straight, gay, tourist, octogenarian, and infant—turns his or her head and says, "Oh!" What is it about that face? What magic do the beautiful have that most of us lack? Even poets struggle to find the words. Emily Dickinson simply said, "Beauty is not caused. It is."

Where poets rhapsodize, scientists analyze—neuroscientists, psychologists, and anthropologists have all taken a stab at deconstructing facial beauty. Overall, they've focused on three measures: averageness (how closely the size and shape of facial features match the average), symmetry (how closely the two sides of the face match), and sexual dimorphism (how feminine or masculine the face appears). We're only talking about facial shape and features here, not age, expression, or complexion.

You might think the first one, averageness, seems odd. By definition, isn't average just average? But most of us don't have average features. When compared to the average, your eyes may

Test Your Facial Symmetry

Unless you're a model or just look like one, chances are you're at least slightly asymmetrical. Don't worry, most of us have faces that are slightly off—one jagged eyebrow fixed higher than the other, or a cockeyed smile. If you have Adobe Photoshop, you can make a quick symmetrical version of your face.

1. Open a passport-type photo of yourself.
2. Using the Rectangular Marquee Tool, highlight and select one-half of your face.
3. Right-click the selection and choose the Layer via Copy option from the menu. This will copy your selection.
4. In the menu bar, go to Edit > Transform > Flip Horizontal. This will create a mirror image of the copied selection.
5. Use the Move Tool to drag the selection to match the other side of your face until you have a whole face. Voilà—a perfectly symmetrical you.
6. Repeat for the other half of your face.

You might be surprised when others think the two right sides of your face are more attractive and resemble you more than the two left sides. That's because the right side of your face leaves a stronger impression in the eyes of your beholders. When you and another person are standing face-to-face, the other person's left eye looks directly at the right side of your face. The left eye is controlled by the right hemisphere of the brain, which processes faces and emotions faster than the left hemisphere, which controls the right eye. Some studies have also found that the two halves of your face reveal different parts of your personality, and that the right side of the face is more powerful while the left is more moody and expressive (see page 181).

be too wide or close-set, your eyebrows uneven, or your nose too sharp. When a batch of faces is "averaged" to make a computer-generated composite, judges rate the composite as more attractive than any one of the faces that constitute it. The more faces blended in the composite, the more attractive the result.

Blending races helps, too. Psychologist Gillian Rhodes asked Asians (Japanese) and Caucasians to rate the attractiveness of male and female faces. Subjects from both cultures thought mixed-race Eurasian composites were much more attractive and healthier-looking than composites of all-white or all-Asian faces. Double-checking her results, Rhodes did another study using the faces of actual Eurasian people, not composites, and reached the same conclusion. The biracial faces were judged better-looking.

Maybe that's why so many models come from Brazil, where so many people are biracial or multiracial.

So what draws us all to the middle? Researchers have several theories. For one, familiarity breeds attraction—we learn to identify patterns in the faces we see, and calibrate our perceptions to match these known patterns. Averaging all the faces we've seen, medium proportions would be more familiar to most of us than distinctive features such as potato noses, wide-set eyes, underbites, and chipmunk cheeks. (However, if you grew up around people with distinctive features, you'd probably find them more attractive than someone who did not.) Distinctive and unattractive features may be telltale signs of undesirable recessive genes. Looking at portraits of the inbred Habsburgs, you can see how members of the ruling house of Europe shared the same DNA to the extent that their looks and health suffered—it shows up in their protruding lower lips, misshapen noses, and doorknocker mandibles. Poor Charles II had a jaw so deformed that he could not chew.

Even an infant might turn her nose up at the Habsburgs, according to studies that suggest that "beauty detectors" are hardwired in our brains. Remarkably, babies who have had very little previous exposure to people have the same facial preferences as adults. Infants as young as one day old, when exposed simultaneously to beautiful and unattractive faces, consistently gaze longer at the attractive faces. The neural mechanism that enables babies to distinguish beautiful from beautiless is unknown, but it is widely agreed that it exists. People from different cultures also generally agree on what faces are hot or not.

Beauty isn't created equal, as you might expect. There's beauty and there's *bedazzling* beauty. The most striking faces are close to the average but with certain optimal "tweaks." Evolutionary psychologist David Perrett demonstrated this by taking a composite of an attractive female face and selectively modifying her features, gifting her with higher cheekbones, larger eyes, and shorter distances between her mouth and chin and nose and mouth. This tweaked composite won the beauty contest over the attractive averaged composite in the way that supermodels trump catalog models. It turns out that atypical features can enhance a person's looks but only in the right place on the right face.

Symmetry, the second measure of beauty, can make or break the equation. Look at actress Gwyneth Paltrow for an example of a beautiful but slightly atypical face. Her mouth is wider than average, and so is the space between her eyes. On another person these distinctive features might not be so stunning, but Gwyneth's face happens to be perfectly symmetrical. This is also true of hotties such as Denzel Washington, Kate Moss, Christy Turlington, and Cindy Crawford (minus the mole).

Not all beautiful faces are symmetrical and not all symmetrical faces are beautiful, but symmetry often plays a role in attraction. Like averageness, symmetry suggests developmental stability. If you grow up with symmetrical features—despite risk of disease, genetic mutations, starvation, pollution, and parasites—there's a better chance you're fit and healthy and your body can ward off infection. Researchers at the University of New Mexico measured the chin length, jaws, lip width, eye width, and height of more than four hundred men and women to determine their facial symmetry. Comparing the results against each participant's health records, they found that people with the most symmetrical features were healthier (i.e., had shorter and fewer respiratory infections and took fewer antibiotics).

Masculinity or femininity (sexual dimorphism) is the third measure of attractiveness. In men, the hormone testosterone is behind prominent jawlines and cheekbones, thicker brow ridges, larger noses, smaller eyes, thinner lips, facial hair, and a relatively long lower half of the face. Women are attracted to rugged, masculine faces because they signal strong immune systems and, potentially, high fertility and social status. Strikingly, women's levels of the hormone estrogen influence how attracted they are to masculine faces. The higher a woman's estrogen level, the more she is attracted to cues of masculinity (see page 124). Estrogen is also behind the beauty of female faces. It plumps out women's lips and skin and produces smaller and pointier chins, smaller noses, rounder cheekbones, eyebrows high above the eyes, and a bottom of the face that is narrower than the top half.

Anthropologist Donald Symons, who in the 1970s first proposed that average faces are beautiful faces, said that we all have "beauty detection" devices in our heads, and it appears he was right. We calculate averageness, symmetry, and sexual dimorphism as easily as Newton calculated numbers. But remember: These three rules only represent physical attractiveness in a general way. Researchers haven't been able to measure the beauty in a person's eyes or the exquisiteness of an expression. For that greater truth you still need poetry.

Beauty is truth, truth beauty.
—Keats

How long does it take to decide if a person is hot?

To find out exactly how quickly we can tell if a person is hot or not, neuroscientists Ingrid Olson and Christy Marshuetz devised a sneaky experiment. They exposed men and women to a series of pre-rated faces, some gorgeous and others homely, and asked them to rate their appearance. The twist was that the faces flickered on the screen for only thirteen milliseconds—a flash so fast that the exasperated viewers swore they didn't see anything. Yet when forced to rate the faces they thought they didn't see, the judges were uncannily accurate. Without knowing why, they gave good-looking faces significantly higher scores than unattractive ones.

The fascinating implication here is that beauty is perceived subconsciously. It's not as if the subjects had much time to meditate on anyone's hotness—they weren't even aware of seeing a face. To a great extent, first impressions of people's looks are less about choice and culture and cultivated tastes, and more about something deeper and universal. Judging attractiveness seems to

Consider Love at Second Sight, Too

Your instincts might draw you to attractive faces within milliseconds of seeing them, but that doesn't mean perceptions are fixed like butterflies under glass. Depending upon your personal experience with a person, beauty can turn ugly, and ugly can become beautiful.

Evidence for this is found in a study by Kevin Kniffin, an anthropologist at the University of Wisconsin–Madison, and David Sloan Wilson, an evolutionary biologist at Binghamton University. They found that people who are liked and respected are more likely to be considered better-looking by the people who know them than by strangers, and people who aren't liked are more likely to be considered less attractive by the people who know them. When people judged one another's looks before and after they knew each other personally, individuals who were initially judged as plain were given significantly higher ratings months later if they were well liked (considered cooperative, dependable, brave, hard-working, kind, etc.). The opposite happened with those who were not well liked. Evidently, we blend the channels of like/respect and physical attraction as we retune our perceptions. That means if you're looking for a long-term relationship that's based on more than raw sex appeal, you might want to give yourself (and him) a chance to fall in love at second sight.

happen just as automatically and matter-of-factly as judging identity, gender, age, and expression.

When you see an attractive face, reward centers of your brain known as the nucleus accumbens and orbitofrontal cortex are stimulated, as well as the amygdala, which captures expression. You also have a specialized cortical network known as the fusiform facial area, which, in a glance, may process a person's whole face: its contour, configuration, and features such as eyes, nose, and lips. It takes coordination between these areas of the brain and others, including the temporal and occipital lobes of the right hemisphere, to form a complete impression of a person's appearance. While some parts of your brain can capture a face in thirteen milliseconds, up to two hundred milliseconds may actually tick by before other parts process the face and a perception emerges on the screen of your consciousness.

Even so, you're faster than you think.

Are you more attracted to people who look like you?

To tackle this tough question, Lisa DeBruine, a psychologist at the Face Research Lab at the University of Aberdeen, photographed the faces of 150 young heterosexual male and female subjects.

Then she did something ingenious with their photos. Using computer graphics software, she blended each subject's facial features with that of an opposite-sex composite. Basically, she made a male version of every woman and a female version of every man. The resemblance was discernible, but none of her subjects consciously detected the manipulation.

DeBruine then showed each subject nine composite faces, including the one that resembled his or her own, paired the faces in thirty-six different combinations, and asked three questions: (1) Which face seems more trustworthy? (2) Which is more attractive for a long-term relationship (marriage or similar)? (3) Which is more attractive for a short-term relationship (an affair or a one-night stand)?

It turned out that men and women gave the faces that looked like themselves the highest ratings for trustworthiness. Yes, most said, they would consider this self-resembling face for a long-term relationship such as marriage. But when they rated the face for a short-term fling, it bombed. That oddly familiar, trustworthy face was marriage-worthy—but not lust-worthy.

DeBruine's study suggests that attraction depends very much on what you're looking for—is it a hot date or a lifelong love? For a fling, you're less likely to be physically attracted to faces that resemble your own; subconsciously, you prefer genetic diversity or the best-looking specimen of the opposite sex. But for long-term relationships, you might desire, or at least have a higher tolerance for, someone who looks familiar. That's because you're looking for other qualities, too—trust, shared values and interests, and maybe someone whose personality reminds you a little of your own. For each of us, the ideal partner has a unique blend of comfort and sex appeal. Too bad we can't create our own real-life composites.

Do women choose husbands who look like their fathers?

The preference for partners who resemble one's parents, especially the opposite-sex parent, is known as *sexual imprinting*. It happens among other animals all the time. Male zebra finches go wild for females with the same markings as their mothers. Goats raised by a sheep prefer to mate with sheep, and sheep raised by a goat prefer to mate with goats. Don't worry, no one's hot for Mom or Pop—but there's evidence that sexual imprinting leaves its mark on humans, too, especially in long-term relationships.

Studies on sexual imprinting have found a loose association between the looks of people's parents and spouses. In a study at the University of Texas at Austin, kids from mixed-race marriages were more likely to marry a person who was the same race as the opposite-sex parent than the same-sex parent. Another study found that the best predictor of the eye color of a woman's partner is her dad's eye color. For men, the hair and eye color of his mom is the single best predictor of his partner's hair and eye color. (Note: A guy might be more likely to marry a dark-eyed brunette if his mother is one.) However, there is less evidence of an Oedipus complex among men than the reverse, an Electra complex among women.

Anthropologists at Durham University in England and the University of Wroclaw in Poland asked forty-nine women to look at photos of fifteen men and rate each one for his desirability in short-term and long-term relationships. When the experimenters measured fifteen facial proportions in photos of the women's fathers—including face height/face width and brow height/face height—they found that for long-term relationships, the women were more attracted to men with their dad's facial proportions. A woman who is a daddy's girl, or was one when she was young, is more likely to go for a guy who resembles him. Women who didn't get along with Dad, or whose fathers weren't around when they were growing up, didn't have a preference.

Proving that women's bias goes beyond men who look like themselves, a Hungarian team studied married women and their adoptive fathers. They found a resemblance between photos of the women's fathers and husbands based on how the adoptive fathers looked when the women were between two and eight years old, the prime age range of sexual imprinting. Again, the more warmth between daughter and adoptive Dad, the more likely she was to pick guys who looked like him when he was younger.

Women's tendency to marry men who resemble Dad, if Dad is loving, adds to the increasing evidence that women lean toward the familiar and the positive for long-term relationships. Sexual imprinting may be a by-product of the way we learn from our parents, or a general preference for what we think is typical in the opposite sex. We may be modeling our marriage on Mom's, or unconsciously deciding that since Dad is a good parent, then a man who looks like him will be, too. It's also possible that we link our opposite-sex parent's personality characteristics with physical ones (for example, Dad's thick eyebrows and strong jaw as a cue of dominance) and unconsciously seek these traits in a significant other. But remember, lest your Freudian imagination runs wild: the attraction is limited to general resemblances.

Can you tell if a man has daddy potential from his face alone?

Yes, chances are good that you can tell at a glance if a man is kid-friendly. A team of researchers from the University of California at Santa Barbara, led by psychologist James Roney, took photos of thirty-nine heterosexual men ages eighteen to thirty-three, tested their saliva for testosterone, and assessed their interest in infants. Then they asked thirty women to rate each man's headshot for qualities such as physical attractiveness and masculinity, fondness for children, and suitability in a short- or long-term relationship.

With uncanny accuracy, and only a still photograph to go by, nearly 70 percent of the women were able to predict which men had papa potential. Their cue? It could have been the shadow of a smile on some of the men's faces, even though they were instructed to keep their expressions neutral. Or the women may have unconsciously picked up an essence of something kinder, softer, gentler, and, well, paternal in their faces. These were the men women said they'd prefer for long-term relationships. The marrying kind.

The women were also asked to rate the men's masculinity. Again with striking precision, they were able to tell which of the men had the highest testosterone levels. Strong jawlines (a broad bottom half of the face) and facial hair were two big giveaways. High-testosterone guys were the ones women thought were hottest and the most desirable for a fling or one-night stand. They were also less likely to be child-friendly. However, to be fair, this wasn't always true. Some masculine-looking men in the study were accurately judged as having paternal proclivities. Yes, ladies: there are guys who are both very manly and kid-friendly. They might not be the norm, but they're out there.

From an evolutionary perspective, it makes sense that women are so good at telling which men are daddy material. Moms benefit from mates who invest in their children and help raise them. As the saying goes, any man can be a father, but it takes someone special to be a dad.

How might your mom or dad's age influence your attraction to older faces?

If your parents had you when they were older than thirty, you've spent your entire life gazing into faces that are maybe a wee wrinkled, a little lined. Mom and Dad might have been a little older than your friends' parents, and so perhaps were your aunties and uncles and family friends. To you, aging faces are familiar. And what is familiar may be more tolerable (and even attractive).

In a study at the University of St. Andrews in Scotland, more than eighty men and women in their twenties were shown composite images of men's and women's faces. Some of the faces were specially manipulated to show wrinkles, lines, bags, coarse skin, and other signs of aging. The researchers, evolutionary psychologist David Perrett and his colleagues, asked the subjects to rate each face's attractiveness in the context of a short- or long-term relationship. They also asked the participants to record their parents' ages.

Overall, women were kinder than men in their ratings, particularly when they looked at aging faces in the light of a long-term relationship. From an evolutionary perspective this makes sense: for marriage, women often care more about men's companionship and status than raw sex appeal, so they often prefer older men. However, Mom and Dad's ages played a significant role in determining how tolerant a gal would get. Women with younger parents downgraded the appearance of aging faces more than did women with older parents, and were unlikely to consider them for a fling.

Men's ratings also depended upon the context of a relationship. For short-term relationships, they gave low attractiveness ratings to older-looking faces. No surprise here—men are biased toward youthful-looking women with childbearing years ahead, and they generally marry women who are younger. However, intriguingly, if a guy's mother was over thirty when he was born, he was likely to be more tolerant of aging in women's faces in the context of a long-term relationship. Only the mother's age at his birth, not the father's, influenced a man's acceptance of older-looking women's faces. (This means if you're trying to gauge a man's tolerance to aging faces, it doesn't hurt to ask him about how old Mom was when he was born.) Further research will reveal if men with older moms more often marry older women. There's evidence that women with older dads more often marry older men.

The psychologists speculate that these biases are due to sexual imprinting—the preference for mates who resemble one's parental figures, especially the opposite-sex parent. Your parents are familiar to you, and you may value familiarity more than sex appeal in long-term relationships. Men and women with older parents might associate aging faces with traits they've learned to value: maturity, dependability, honesty, competence, and so on. With those priorities, they'd make their parents proud.

Why do blue-eyed men prefer blue-eyed women?

Bruno Laeng, a psychologist at the University of Tromsø in Norway, recruited nearly four hundred men and women and asked them to identify the shade of their partner's eyes. Only one subset of subjects showed a clear-cut preference: men with blue eyes. Nearly 70 percent of the blue-eyed guys had blue-eyed girlfriends, whereas men and women with any other eye color, as well as blue-eyed women, couldn't care less about the eye color of their loved ones. In a second experiment, Laeng and his colleagues altered the eye color of women in photos and presented two versions to men: one with blue eyes and the other with the women's natural shade. Again, only men with blue eyes showed a strong bias toward the blue-eyed version.

According to Laeng, blue-eyed men are picky for a reason. They might see a woman's soul when gazing into her eyes, but they're also looking at her genes. Blue eyes are recessive, meaning that eyes only get to be blue in the absence of dominant brown-eyed gene variants, and are therefore rarer than other eye colors. If a couple are both blue-eyed, their child will have blue eyes. It's virtually impossible for two blue-eyed parents to have a brown-eyed child, but if the wife cheats on the husband and conceives a baby with a guy with any other eye color, the child is much less likely to have blue eyes. Seen this way, blue eyes are a simple and predictable paternity (or fidelity) test.

Now, it's not as if every blue-eyed guy you meet looks at your eyes and consciously thinks about his paternal legacy. The researchers say the bias is subconscious, and exactly how men learn or intuit it is unclear, but they're right to have a niggling concern about infidelity. Approximately 3–4 percent—although some studies say up to 10 percent—of all babies born in the United States and Europe have fathers who are presumed to be their biological dads but aren't.

Of course, the strategy is not foolproof. If a blue-eyed wife has a liaison with another blue-eyed guy, her blue-eyed husband has lost his advantage. In addition, Caucasian babies are frequently born with a neutral (often blue-gray) eye color by default, with the eyes darkening after a year or so; this could be nature's way of protecting infants from suspicious dads. But when he discovers the truth, the blue-eyed guy might start seeing red.

CHAPTER 2
Following Your Nose

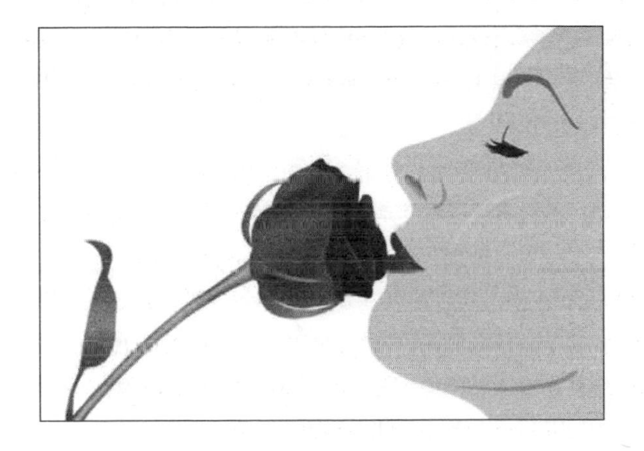

Oh strong-ridged and deeply hollowed
nose of mine! what will you not be smelling? . . .
Must you taste everything?
Must you know everything?
Must you have a part in everything?

—William Carlos Williams, from "Smell"

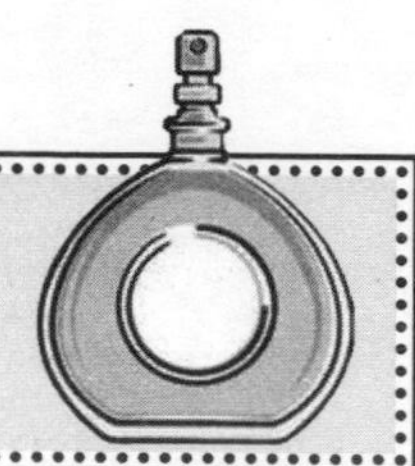

What are pheromones, and do they exist in humans?

Pheromones are chemical signals that, when released in the air, trigger specific behaviors in other members of the same species. Territorial pheromones in dog pee say, "This spot is mine!" and other dogs sniff and stay away. Alarm pheromones in aphid secretions say, "We're under attack!" and other aphids mount a defense. Sex pheromones say, "I'm horny!" or "Assume the position!" and a whiff of them is all it takes for animals to hump. Pheromones are nature's aphrodisiacs, and Mother Nature is a generous dealer. Bees use them, birds use them. So do fleas, chimpanzees, and kangaroos. So why wouldn't we?

The controversy over human pheromones is less over whether they exist than what they *do*. While chemical signals are perfectly acceptable in the simple, bottom-baring sex lives of monkeys and pigs, people are different. Human sexuality is subtle (well, sometimes), which makes the verdict complicated. But change is in the air.

We know other mammals detect pheromones with their vomeronasal organ (VNO), two pits located in the back of the nasal septum between the nose and mouth. Only recently have scientists discovered real evidence of a VNO in human adults. Skeptics point out that we don't have proof that our VNOs are still connected to our brains—the link may have been lost or diminished as our species evolved. Fortunately, even if our VNO isn't what it used to be, it's not the only pathway pheromones might use to get to our brains. There is mounting evidence that humans and other animals can also process pheromones the same way we process other smells, through the main olfactory system. Pheromones could trigger our sex drive by traveling the neural pathway that connects the nose to the hypothalamus, a region of the brain that initiates the release of sex hormones and fuels erotic feelings and sensations. Remarkably, people who have lost their sense of smell often also report a loss of libido.

Pheromones may be in body fluids: sweat, saliva, semen, vaginal fluids, breast milk, and blood. You'd inhale them in the air as you would a perfume, or by kissing. If you don't like the way your beau smells, you might find yourself dumping him no matter how good-looking, charming, or successful he might be. The scent of men's sweat might also make you happier, hornier, and more focused. Pheromones may even play a role in helping you choose a partner and increase your chances of getting pregnant. (More on all these to follow.) The effects of pheromones are subtle, but proof that they exist is real and growing.

Psychobiologist Charles Wysocki likens pheromones to the active ingredients in traditional medicinal teas—they're difficult to isolate even though people swear by the results. Researchers be lieve that one major source of human pheromones is the immune system's major histocompatibility complex (MHC). Whether you're attracted to or repulsed by a man's body odor may depend strongly on your respective immune systems. Other potential pheromones are testosterone derivatives such as androstadienone and androstenone found in sweat. Female pheromones may be related to the hormones estrogen or progesterone and are present in sweat and vaginal fluids.

Chemical signals may influence our love lives, but so do a nearly infinite number of other things that tug us this way and that way. The complexity is a good thing. It's all that separates our sex lives from mere respiration and fornication.

Why do some men smell better to you than others?

Admit it—you've loved a lover's smell so much that you would sniff his dirty shirts when he wasn't around (we're talking natural body odor only, not cologne or other smells). At least once you've decided not to change the sheets so you could wallow in his aromatic embrace for another night. If so, it's a good sign. A man's natural scent probably has something to do with why you're with him in the first place.

T-shirt sniffing has also been done under laboratory-controlled conditions with consistent results. Women sniff T-shirts that have been worn for several consecutive days by anonymous men and are asked to identify which ones smell (relatively) appealing. The guys aren't allowed to wear deodorant or cologne; use scented soap; eat food that contains garlic, cumin, or curry; or

Keep Away the Pheromone Spray

"Release your lover's raw animal sex drive with pheromone concentrate!" This and similar boasts about pheromones are backed up only by industry-driven research, not objective science. Truth be told, there is no hard evidence that men or women have responded to any pheromone in the way that lonely hearts hope. The problem is that pheromones aren't that simple, nor are people. Humans don't respond predictably to pheromones. Nor are all pheromones universally attractive; some are sexy only because they signal genetic compatibility, which varies from person to person. Plus, there isn't any proof that the compounds in commercial sprays are even natural human products. Some contain animal urine. If you want to try to unleash your stud's "raw animal sex drive," it might behoove you to lay off the horse pee and spend the cash on a bottle of good wine.

smoke, have sex, or drink alcohol. The smell of their sweat is as pure and untainted as possible.

In study after smelly T-shirt study, a pattern emerges. Women prefer the body odors of guys who have major histocompatibility complex (MHC) gene variants that are mostly different from their own. The MHC is a cluster of immune system genes responsible for detecting and identifying bacteria and viruses that invade the body. You inherit them from both parents. The diversity of MHC variants in the human population is a reason why people vary in their immunity to various bugs, from common colds to plagues, superflus, and other deadly scourges. While organ recipients seek donors who share as many MHC variants as possible, women are attracted to guys who share only a few. Women aren't alone in preferring mates who have mostly dissimilar MHCs—so do mice, birds, fish, and many other animals.

Like most other aspects of attraction, your bias has to do with reproduction and survival of our species. If you have children with a man whose MHC variants are mostly unlike your own, your kids may inherit a more diverse MHC and stronger immune systems that identify and destroy a greater range of bacteria and viruses. If your partner has MHC genes that are very similar to your own, your children might not be as healthy.

You might be wondering how you smell MHC in a guy's sweat. Researchers speculate that MHC genes code for specific proteins that circulate in the bloodstream. These proteins bind to odorants in concentrations that depend upon a person's unique MHC. The odorants, in turn, ooze out of sweat glands in the armpits and genital areas. When you snuggle with your guy, you may be whiffing those odorants directly, along with other odors in his sweat. As gross as it sounds, the microbes thriving in sweat glands also break down the MHC-related proteins, and the resulting body odor is at least partly bacterial by-product. The musky stench in male sweat comes not from MHC but from the testosterone derivative androstenone, which is also a potential pheromone. (See page 38.)

Don't worry that guys might turn up their noses at your MHC the way you do at theirs. Women are much more sensitive than men to smells. (About 60 percent of women can identify their personal odor compared to less than 6 percent of men.) While unappealing body odor is a deal breaker for many women, many guys are oblivious to it.

Seen through an evolutionary lens, it makes sense that women are better than men at detecting MHC odors of potential mates. Women have historically put more time, energy, and risk into parenting than men have, so we're the choosier (and more sensitive) sex. With so much invested in our children, we've evolved to sniff out every subtle cue available that helps us assess the health and compatibility of our mates. A body odor that indicates a genetic mismatch may be a big red flag—your body telling you he's wrong for you on a base biological level. (It doesn't matter if you actually want children; we're wired this way.)

So nuzzle! Dig your nose into your lover's hair, his neck, his chest, his armpit, his belly button, and so on. Inhale and let the chemicals hit your brain. Then use your senses.

Go Off the Pill Before You Say "I Do"

A strange thing happens when scientists ask women taking hormonal contraceptives (birth control pills, aka the Pill, or Depo-Provera) to rate the odors of male T-shirt wearers. Unlike women who aren't on the Pill, they often prefer the smells of men whose MHCs are *similar* to their own. Equally disturbing, the Pill might diminish the "sexy signals"—subtle changes in your looks and behavior—that attract men. That's because hormones such as estrogen and testosterone that normally spike during ovulation actually flatline when you're on hormonal contraceptives (see page 120). Researchers are unsure why the Pill reverses women's usual preference for men with MHC-dissimilar genes, but it's evident that it has something to do with the lack of hormonal fluctuations. Pill-takers do not ovulate.

As convenient as oral contraceptives are, they may seriously compromise your biological instincts. Imagine if you marry a man, go off the Pill, and decide he doesn't turn you on anymore. If you meet someone when you're on the Pill, you might want to go off it to make sure you're still attracted to him before you commit.

Why might you cheat on a man if you don't like his smell?

There's no infidelity gene that drives you into the muscular embrace of a lover. However, there is astonishing evidence that a couple's genes, when compared to determine how similar they are, may *predict* whether the woman will be unfaithful to her partner. The genes are the major histocompatibility complex (MHC), which is involved in immune function and affects the production of body odors. As we now know, women prefer the smells of men whose MHC variants are mostly unlike their own. This means if you don't like your partner's natural body odor, you might be too genetically similar. It's your body's way of telling you that you want a man who's mostly genetically *dissimilar*.

Exploring the connection between MHC and infidelity, evolutionary psychologist Steven Gangestad, doctoral researcher Christine Garver-Apgar, and their colleagues at the University of New Mexico asked about fifty long-term couples to answer questions about their relationships. How attracted were they to their mates? How satisfying were their sex lives? How much did they fantasize about other lovers? Have they had any affairs? When and for how often?

A shocking pattern emerged. While the average couple has 20 percent of their MHC variants in common, some couples have more. The more MHC variants a woman shared with her partner:

- The less sexually attracted she was to him
- The more she rejected his sexual advances
- The more she fantasized about other guys
- The less aroused she felt when having sex with him

Call it bad chemistry.

Unsurprisingly, some of the women found solace in the arms of other men. Then, when they had sex with their MHC-similar partners, they reported fewer orgasms than with their lovers. If a woman is having an affair and climaxes with her lover instead of her partner, the lover gets the reproductive edge (see page 212 to learn why).

There are known medical risks in having a baby with a man whose MHC genes are too similar to your own. Homozygous babies (offspring that inherit the same gene variants from both parents) are more frequently miscarried, and even in successful pregnancies they tend to be sickly and/or underweight. Heterozygous babies (offspring that inherit different gene variants from both parents) have stronger immune systems that can resist a greater range of assaults, including parasites and hepatitis viruses. And a bonus: your heterozygous kid may also be better-looking and have better skin, as found in studies that compared the physical attractiveness of people who have a diversity of MHC variants and those who don't.

So does following your nose lead you to infidelity? From a purely evolutionary perspective, yes. Our ancestral mothers may have desired a hunky foreigner's genes for her children (MHC-

Test Your MHC Compatibility

There's already at least one dating service that finds matches for singles using MHC compatibility testing. Is it an omen of the future? Imagine going to the drugstore and picking out an MHC compatibility test in the way you would a pregnancy test. It'll herald a whole new era of "genetic horoscopes" that lay out your marital destiny. Seriously, if marriage counselors and sex therapists tested their clients' MHC compatibility, they might develop a new take on relationships. For better or for worse, MHC compatibility is not commonly tested, although it's possible for a lab to compare the proportion of MHC variants you and your lover have in common. All that's required is a scraping of cells from the inside of the mouth. If that's too much, you might want to challenge your beau to abstain from showering and eating spicy foods for a few days. Once he's pungent, do your own smelly T-shirt test.

dissimilar) but also appreciated the safety and security of kin (MHC-similar). This sort of push and pull also shows up in studies that suggest women are attracted to a "foreign" hunk's body odor for sex alone, but for long-term relationships favor facial features of men who may be somewhat genetically similar. Perhaps it was a good strategy for ancestral woman to occasionally have flings with outsiders, or at least find men in their communities who were maximally MHC-dissimilar. The latter is true of the Hutterites, a socially isolated religious community living in the Midwest. When geneticist Carole Ober took DNA samples of couples, she found that women subconsciously pick mates whose MHCs are as different as possible from their own within the limits of a closed community.

Many men you meet, especially if you live in a diverse enough area, are likely to be a happy medium for you—not too similar and not too dissimilar. But what do you do if you've fallen in love with a man whose odorprint you can't bear? It's up to you. You can follow your nose, or you can follow your heart. Clearly, some women have followed both.

Why might the smell of men's sweat brighten your mood and senses?

What do men's locker rooms have in common with your sex life? The answer is the scent of androstenol, a derivative of testosterone that is produced in men's testes and present in their sweat, skin, semen, and blood. (It's also found in truffles and is the reason why pigs sniff them out.) To most people, androstenol, androstadienone (*an-dro-sta-DIE-en-own*), and the related steroid androstenone (*an-dro-STEEN-one*), have the reek of pungent musk. To others they don't have much odor, and to a much smaller percentage of us they smell like sweet vanilla. How you perceive them depends on your variant of an odor receptor gene. (Don't confuse these chemicals with those linked to the immune system's MHC genes.)

The surprising thing about androstadienone is that it has been found to change women's body chemistry. In a study at the University of California at Berkeley, neuroscientist Claire Wyart and her colleagues asked nearly fifty women to take twenty sniffs of synthetic androstadienone while watching movie clips. In a separate session on a different day, she asked the same women to watch movies and take twenty sniffs of yeast. The two sessions had markedly different outcomes. After sniffing yeast, nothing special happened. But after sniffing androstadienone, women reported higher levels of sexual arousal and more positive emotions. Strikingly, the women's hormone tests also showed increased levels of cortisol, which is associated with arousal, better memory, and increased attention. Within fifteen minutes of the first whiff of androstadienone, cortisol levels were significantly higher than at baseline or when smelling yeast, and their blood pressure, heart rate, and breathing increased. Women were hopped up on the hormone for up to an hour after the experiment.

There's more and more evidence that androstadienone affects us in subtle yet substantial ways. A study at the University of Pennsylvania found that women feel more relaxed after inhaling extracts of male underarm sweat. At the University of Chicago, researchers concluded that the hormone can boost women's desire and mood after exposure to only a tiny amount, and a PET scan showed increased activity in brain regions associated with vision, emotion, and attention when executing a visual task. A Berkeley study found that sniffing androstadienone enhanced women's moods when watching film clips. They were more aroused than a control group when viewing an erotic film, and maintained a relatively upbeat mood after watching a sad film, whereas men's moods darkened. (Male sweat doesn't have any positive effect on heterosexual men.) The compound also impaired women's memory of negative events in the sad film.

Most fascinating of all, women judge the odor of androstenone (a related compound) as most pleasant during the fertile phase of their menstrual cycles. The hormone may have a subtle mind-altering effect on ovulating women, lifting our moods to ease the way to sex and pregnancy. If you're near ovulation, the presence of a high-androstenone (testosterone) hunk might even hasten the release of the egg. One study found that women exposed to male sweat experienced an increase in the frequency of luteinizing hormone (LH) pulses. The more LH your body secretes, the sooner you ovulate. By coaxing your ovaries to release an egg, your lover may be unconsciously increasing his chances of getting you pregnant.

At least he'll probably be a good-looking daddy. According to a study at the University of New Mexico, a symmetrical man's sweat smells better to women than that of a lopsided fellow. Androstadienone and its cousin compounds are derivatives of testosterone, so on a biological level it makes sense that the sweaty smells women prefer are from hunkier men.

You might wonder how obvious these effects are in real life. The answer is that you're probably oblivious to them. The androstadienone in male sweat may enliven your senses and lift your mood enough for a sexual encounter, but it's not like the potent punch the steroid delivers to other animals. If a female pig catches a whiff of boar androstenone, she promptly perks up and assumes a wiggling, butt-up, legs-spread mating stance. In humans, these compounds don't seem to *create* a mood as much as *modulate* it. When watching movies, happy reactions simply become happier, sexy thoughts become sexier. Or when being near a man, but not a woman (if you're straight), your mood may shift and you become more attentive or feel a libidinous shiver. Androstadienone may do for your sex life what caffeine does for your mental activity—enhance your mood and alertness. But only in the right context.

Researchers know that every person has a genetically determined sensitivity to androstadienone: some of us are hypersensitive to concentrations in sweat, and others don't seem to detect it at all. In real life, the effects may be complicated by other signals, such as a man's MHC compatibility, diet, and body hygiene.

A friend of mine admitted that when she works out at the gym, she always scans the room for a good-looking sweaty guy and then chooses the station next to him. Not that she's looking for a date; she's happily married. She says she secretly likes the pungent savor. I asked her what it does for her, and she shrugged. "I just call it my workout high."

Join a Subculture

If you're curious about pheromone experimentation, do a Google search for bloggers who dabble with synthetic androstadienone and other testosterone derivatives. You'll find an enthusiastic subculture out there comprised mostly of geeky guys who do DIY experiments, mixing and matching methodically to find the pheromonal sweet spot that arouses women or makes them friendlier. (Many pheromone-type chemicals are available from online chemical suppliers.) It's all interesting and amusing, if not exactly scientific. There are the successes—stories from guys who swear that pretty women gaze at them longer and act friendlier. Then there are the disappointments—like the one from an experimenter who complained bitterly about how women struck up conversations with other men on the subway, but not with him, even though he was the source of the pheromone that was making women so friendly. Then again, who knows?

Does women's body odor have any effect on men?

Maybe. We know that body odors are probable cues to good genes and genetic compatibility, and that women are much more sensitive than men to them. But here's a fact that might astonish you: Not all men are senseless.

In an experiment at the University of Texas at Austin, evolutionary psychologists Devendra Singh and Matthew Bronstad recruited nineteen female volunteers and gave each of them two T-shirts. For three consecutive nights during their ovulatory phases—the time of the cycle when they were most likely to get pregnant—the women wore one of the T-shirts, and for three consecutive nights during a nonfertile phase they wore the other. The researchers then asked fifty-two men to sniff each pair of T-shirts and choose which smelled better.

Without knowing why, the majority of the men found that the T-shirt women wore in their fertile phase smelled significantly more pleasant than the T-shirt worn at an infertile phase of the menstrual cycle. Even after the T-shirts were left at room temperature for a week, the men judged the fertile body odors as more attractive. Of course, the guys didn't *consciously* know that the smells they preferred were linked to women's fertility. (However, the findings are significant only when researchers compare each woman's odor at various points in her individual cycle, rather than comparing the odor of one set of women who are in their fertile window versus another set of women who are not. Other studies in the Czech Republic and Finland had the same outcome.)

Sweat is not the only body odor that might betray your fertility status. Vaginal secretions contain copulins, which are fatty acids controlled by the level of estrogen in your body. They fluctuate with your cycle; you produce the most at the fertile midpoint. Several studies have found that copulins smell their most pleasant in the several days before ovulation, suggesting you taste better "down there" at that time. Before our ancestors became bipedal, genital odors were undoubtedly more important than they are now. (My favorite theory on why some men get so aroused by giving oral sex is that they absorb pheromones directly. The same could be true of women who get aroused by giving blow jobs.) Oral contraceptives, meanwhile, eliminate your cyclic scents, and could also change the way you taste when a man goes down on you.

From an evolutionary perspective, any trace of female fertility gives men a reproductive edge. Men may have evolved to be more attracted to women's "fertile smells" as a cue to have more sex at that time of month, and to know when to guard a mate from other males who might impregnate her instead. Not that every man can detect these aromatic shifts. A sensitivity to subtle body odor fluctuations may benefit men in long-term relationships, because only they would be intimate with their mate's smells as they changed throughout her cycle. It might not be any coincidence that, unconsciously, men tend to act more devoted and proprietary when their partners ovulate.

The "fertility odors" that men smell are likely hormonal, but their exact source is unknown. Researchers speculate that they're probably related to estrogen, because estrous urine is a tremendous turn-on in other mammals. Men's ability to detect women's sweat and vaginal odors appears to be weak and easily overwhelmed by other smells such as food, perfume, and pollution. That's a good thing if you think about it. Imagine how traumatized you'd be if every guy knew your fertility status with a casual whiff.

Whether the substances responsible for these body odors can be called pheromones depends upon whether they actually influence men's behavior, and that has yet to be proven. Some men are better than others at detecting and responding to body odors, and it's possible that the ones with good noses get more action when women are most fertile. For now we only know that when it comes to sex, male sensitivity counts.

Can you tell people's sexual orientation by their smell?

You might not be able to smell a sweaty T-shirt and tell if its owner is gay or straight, but there's a good chance your own sexual orientation will color your perception of it.

Consider a study of more than eighty straight and gay subjects recruited by Yolanda Martins and Charles Wysocki at the Monell Chemical Senses Institute in Pennsylvania. The researchers told the men and women to deeply inhale the smells from sweat-embedded pads that had been tucked in the armpits of straight and gay donors. Raw armpit odor contains potential pheromones: androstadienone (in men's sweat) or estrogen-related odors (in women's sweat), as well as MHC-related odors. They asked the subjects to rate the smells, and did not reveal their sources.

It turns out that straights and gays had very different and distinct reactions to the odors. The sweat smells of gay men provoked the strongest love-it-or-hate-it reactions. No shock here: gay men found the body odor of other gay men attractive. But strikingly, *only* gay men found the odor attractive. Everyone else rated gay body odors the most unpleasant of all. To explain, the research suggests that the same MHC genes that affect body odor could also affect a person's production of and sensitivity to testosterone, which in turn may be linked to sexual orientation. Gay men may have one or more unique odor components in their sweat not found in women or other men.

Responses to the other odors were also specific to sexual orientation. Gay men disliked the body odors of straight men and lesbians (but, surprisingly, didn't mind that of straight women). Straight women and men preferred the body odors of other heterosexuals, both male and female, and so did lesbians. The researchers believe that lesbians are more "sex-flexible" than homosexual men, and that genetic influences are significantly stronger in gay men than in lesbians.

Straights and gays also show different brain activity when they smell pheromone-like substances, according to a research group at

Follow Your Nose

Seriously consider what your body is telling you if you don't like the smell of a man's natural body odor. Vivian, a lovely and compassionate woman, married John, a man whose body odor she'd never liked. It didn't repulse her, but it turned her off a little. It was, as she described it, "funky." Viv and John's sex life was formal and uninspired. They had sex only a few times a year, but she blamed herself because in the beginning he had tried but she wasn't feeling it. Unlike with previous boyfriends, she never wanted to dig her nose into John's chest or smell his dirty T-shirts. It wasn't a hygiene problem—he bathed often. After eight years of marriage, John told Vivian that he was gay and had been surreptitiously having sex with other men. When I told her about pheromone studies involving gay men, something clicked. Maybe the odor problem was something else—the food he ate, or perhaps Vivian and John were MHC-similar. But Viv is now one gal who warns women to proceed with caution in a relationship with someone whose smell they don't like.

the Karolinska Institute in Sweden. Neuroscientist Ivanka Savic and her colleagues asked nearly forty gay and straight men and women to sniff high concentrations of synthetic androstadienone (a derivative of testosterone) and estratetraenol (a derivative of estrogen). For purposes of comparison, the sniffers also inhaled odorless air and lavender and cedar oil. After each sniffer took a whiff, his or her brain activity was tracked in a PET scan.

Everyone's brain looked the same when sniffing lavender and cedar oils, which lit up the odor-processing regions more than when breathing odorless air. But something very unusual happened in the sniffers' brains when exposed to the potential pheromones. Depending upon the person's sexual orientation and the substance, the hypothalamus also lit up (although, in the clinical setting of the lab, the subjects didn't necessarily feel sexually aroused). The hypothalamus is known as the "master gland" because it's involved in so many functions, including releasing sex hormones and triggering erotic feelings and fantasies. Your hypothalamus may also play a role in choosing sex partners.

Sure enough, straight women and gay men got all hot and bothered in the hypothalamus when sniffing the male pheromone androstadienone, and straight men's hypothalami fired up when sniffing the female estrogen-like compound. Meanwhile, lesbians showed results that were similar to straight men's—a section of their hypothalami was activated when smelling the estrogen derivatives—but the results were less clear-cut and consistent. Researchers do not know if these biases are due to physical differences in the brains of straights and gays, or if they just process the smells differently due to past experience.

What has emerged from these studies is the theory that we subconsciously sniff out potential mates. It's fascinating, yet questions remain. What are the biological reasons gays and straights have different body odors, and what drives our response to them? Does body odor trigger "gaydar," the ability to detect who's gay? Maybe. Instinct, as the saying goes, is the nose of the mind.

Why might your sex drive pick up around breast-feeding women?

Some pheromones guide us to select certain mates over others. Other pheromones, in their mystical way, seem to gently take us to the next step: baby making. If there are pheromones in breast-feeding secretions, they'd fall into the second category. Incidentally, the idea is not new—there's a long-standing tradition in some cultures for newlywed women to spend time with new mothers to increase their own chances of getting pregnant. But now there's a new spin on the old magic: scientific support that there's something about lactating mothers that really *does* affect other women's sex drives.

The research comes from the Institute for Mind and Biology at the University of Chicago, where a research team instructed ninety childless women to swipe smelly pads under their noses every morning and evening for three months, and to keep a record of their sexual activity and erotic fantasies. Half of the subjects were given pads that had been tucked in lactating mothers' nursing bras and armpits. The other subjects were given pads saturated with an equally smelly placebo. By the second month of the study, the difference between the two groups was significant. Among the women exposed daily to breast-feeding secretions, those with regular sex partners experienced a 24 percent increase in sexual desire as measured on a standard psychological test, and women without partners experienced a 17 percent increase in sexual fantasies. The sexy effect continued through the second half of the women's menstrual cycles, when libido normally declines.

According to the study's authors—biopsychologists Martha McClintock, Suma Jacob, Natasha Spencer, and their colleagues—lactating women emit chemical signals. (McClintock is credited with being the first scientist to introduce the idea of human pheromones.) These signals say, "Get pregnant, too!" and our bodies seem to comply.

Why this happens is unknown, but the answer may be rooted in the evolutionary advantage of "synchronized pregnancy." The theory goes that in ancestral times childbirth was risky and the food supply was inconsistent. Timing was key to survival. Imagine a group of Stone Age women attending to a breast-feeding mother, all the while picking up her chemical signals. Her pheromones subtly indicate that food is plentiful and it's a good time to have sex and reproduce. The attendants would get pregnant and give birth around the same time, and therefore benefit from one another's support. In the event of a mother's death, there'd be other lactating women available to nurse the baby. Babies born within this support system of lactating mothers and other kids could have a better shot at survival in risky environments. Although the theory as applied to humans is controversial, synchronized birthing is a survival basic for other animals.

While the magic ingredient in breast-feeding secretions remains a mystery, it may be related to lactating women's high levels of the hormone progesterone. Progesterone itself may be related to intimacy and bonding. Although it's not directly related to sex drive, it may trigger the production of other sex hormones or affect women's sex lives in other ways. Breast duct secretions, breast milk, or even the baby's saliva (or all of the above) may carry the chemical cue. You'd pick it up in the air, by touching a breast-feeding woman, or by handling something she touched and then putting your hand near your face. If it acts like other alleged pheromones, it stimulates activity in your hypothalamus, which triggers the release of hormones in your bloodstream, which in turn boosts your sex drive.

Might breast-feeding pheromones increase fertility by priming women for pregnancy? It's possible. Unfortunately, no studies have been done yet on women who have direct contact with breast-feeding mothers. So try it yourself by spending a lot of time around a lactating friend and her baby. If you feel a stronger desire to have sex, or have vivid sexual fantasies, it may be some chemical voodoo working to overcome the prospect of dirty diapers and sleepless nights.

What might your perfume reveal about you?

A perfumer might tell you that Chanel No. 5 is the essence of Marilyn Monroe, who loved it so much that it was the only thing she'd wear in bed. Advertising spin, you say, and of course it is. But your fondness for the notes of rose, sandalwood, vetiver, and vanilla in Chanel No. 5 really could reveal something about *your* personal essence—that is, your DNA.

Several years ago, Manfred Milinski, a behavioral ecologist at the Max Planck Institute in Germany, and his colleague Claus Wedekind took DNA samples from more than 130 male and female students. The researchers asked the subjects to sniff common perfume ingredients such as vanilla, patchouli, lilac, sandalwood, and musk, and rate the odors based on how much they liked them and would wear them as a fragrance. Two years later, they asked the same subjects to take the test again. What they discovered is a consistent link between a person's genes and his or her taste in fragrances. Subjects who shared the same variants of immune system genes, the major histocompatibility complex (MHC), tended to like the same scents. For instance, people with a certain MHC variant liked the smells of cardamom, cinnamon, and sandalwood more than people who did not have that variant.

The MHC interests scientists because there is already strong evidence that this cluster of immune system genes produces proteins that are associated with body odor and sexual selection. According to the researchers, fragrances may mimic and amplify a person's natural MHC-related smell, or at least some aspect of it. When you spritz on a scent that contains rose oil, for example, you could unconsciously be publicizing your unique genetic makeup to others. A provocative theory mentioned by the researchers is that wearing a fragrance could be a way of strategically advertising MHC variants that signal resistance against specific infections.

The strongest MHC association was found with the scents of musk, rose, vetiver, and cardamom. These scents have been used for five thousand years, ever since ancient Egyptians invented perfume. Fragrances described in the Bible have essentially the same ingredients as today's designer perfumes. This leads Milinski and Wedekind to wonder if there isn't something very special about the small percentage of the world's plants that have stood the test of time as scents. It could be that these fragrances are the only ones that resonate with human body odors.

You might wonder how exactly your fragrance preferences would reflect your MHC. The researchers don't elaborate, but it could be that people with similar MHCs also have similar sweat and skin chemistries, and certain fragrances combine with or amplify those body chemistries better than others. Further studies may shed more light on the connection between smells and MHC groups. (For instance, I'd love to know why I hate musk, as do many people with a certain MHC variant.)

Milinski and Wedekind could not find a significant connection between a person's MHC and the scents that would turn them on if worn by the opposite sex. That may be because it's difficult to imagine exactly how a fragrance would smell on a lover, or culture and experience may make a clear-cut biological preference difficult. Milinski believes there's some overlap between the ingredients people love in their own perfumes and the ingredients they love on their lovers, just as there's ideally a small overlap in a couple's MHC genes. But for now, putting on airs to attract a mate remains as much an art as a science.

Test Your Fragrance

As with men, you might not be able to tell if a perfume is good for you until you take it home and try it out. And even after you commit, it may change without warning. In fact, some days even your favorite can really stink. Here's why.

First of all, the various "subscents" in a fragrance won't necessarily come out until they have time to interact with your skin. And your skin has its own chemistry, so the combination of scent and skin isn't always predictable. (That's why a good fragrance shop sends customers home with samples.) Spray the perfume on the pulse points of your wrists and behind your knees, where warm blood heats up the scent; in your cleavage, where it'll waft up;

and in your hair, which holds a surprising strength of fragrance. Reapply the next day as well, to make sure you consistently like the smell. Your skin will absorb certain oils and release others.

Just because you like the smell most of the time doesn't mean it works all the time, and that's because certain situations affect your baseline odor. Your MHC genes affect body odor, but many other factors also affect it. Stress is the number one culprit. When you're frazzled, your sweat contains hormones that alter the way you smell, which in turn affects your perfume. Diet is another. When garlic, onions, and curry are oozing from your pores, Love Potion No. 9 smells like something you'd order off a take-out menu. Also affecting the fragrance and your perception of it are the hormonal shifts of puberty, pregnancy, aging, and oral contraceptives.

Can the smell of food or a fragrance be a turn-on?

Maybe, but not because any particular smell is erotic in and of itself. Of the vast array of odors detectable to the human nose—somewhere between ten thousand and forty thousand—not one of them is a universal aphrodisiac. (Sorry, Calvin and Dior.) That's not to say fragrances don't have any effect—they just vary widely in their appeal.

Neuropsychologist Rachel Herz at Brown University has a theory about the impact of odor on the psyche: the emotional connection you have to a smell colors your perception of it. If you have a positive association with a scent, you like it more, and if you have a negative association, you like it less. And if you associate a smell with something sexy, the smell can become sexy.

In one experiment, Herz and her colleagues asked a group of men and women to play a loud, frustrating computer game in a room where a hidden machine pumped out odors that were unfamiliar and unpleasant. The other participants in the study either quietly flipped through magazines in a room permeated with the same odor or played the same game in an odorless room. Subconsciously, their emotions colored their perception of the odor. Participants who played the obnoxious game in the smelly room disliked the odor significantly more than people exposed to it when reading peacefully.

Proving you also like a smell more by association, the researchers pumped a combo of buttered popcorn, rain, and dirt odors into a room where participants played a humorous game and watched a comedy. That group rated the odd odor as significantly more pleasant than did the control groups who weren't having a pleasurable experience while smelling it. (Note: not all smells are purely perceptual. Some are considered universally foul, such as sulfur, decay, and an outhouse on a hot day.)

Odors are processed in your olfactory cortex, the part of your brain that's directly connected to the neural circuits for emotion, memory, and motivation. Odors are unlike images or words in that they often maintain their original association. They can be so precise it's startling. Think about it: have you ever inhaled something that spirits you right back to a certain place and time in your life? (For me, it's the smell of an ex-boyfriend's cologne. Once in a while I catch a whiff of it in a crowd and it transports me in the way the madeleine did Proust.) If the smell of a cologne turns you on, consciously or not, it may remind you of positive sexual experiences you've had in the past, or trigger other associations that put you in the mood. The same can be said for scented candles, incense, potpourri, and other sensory props people employ for romantic purposes.

Even food can be arousing when the right associations are made, as shown in an experiment by psychologist Alan Hirsch at the Smell and Taste Treatment and Research Foundation. Hirsch asked men to inhale the odors of food and perfume as he measured their erections. A combo of pumpkin pie and lavender did the trick; smelling the cozy, domestic odor, men experienced a 40 percent increase in penile blood flow. Other libido raisers were a doughnut-and-licorice blend, the scent of lily of the valley, and junk foods such as cola and buttered popcorn.

Hirsch ventured an evolutionary reason men are turned on by food odors. After a kill or a harvest, our ancestors may have had the best chance of finding a mate and procreating, and therefore males evolved to be sexually primed by food. If that's the case, they're in good company: bonobos and other primates commonly get erections at the sight of a meal. Once sated, sex is next.

Feed His Senses

Cooking smells turn men on; at least that's true of anyone I've ever dated. There's something about food that has a certain sensuous appeal, or perhaps it taps into several primal drives at once. But you have to find out which smells arouse your true love. Cook a meal with him. Open your spice cabinet for him. Create your own concoctions and see if there are any that really excite him. Slather them on your skin and make him taste them. Vanilla, liquor, tomato sauce, oranges, corn on the cob, tortilla chips, jelly doughnuts, cherry pie, peaches, chocolate bars . . .

Is a meat diet a turn-off?

At the height of the meat-only diet craze, many women had a dilemma. They wanted to be supportive of their partners' weight-loss ambitions, but it was getting unbearable—the body odor and the breath, that is. Something about the reek of men on a regimen of steak, chops, and hamburgers interfered with women's desire to cuddle and have sex.

Curious about whether meat consumption really affected attractiveness, Czech anthropologist Jan Havlíček recruited seventeen male volunteers and put roughly half of them on a meat-heavy diet and the other half on a comparable vegetarian diet (for example, risotto with pork versus risotto with vegetables). After two weeks of eating either meat or vegetables, the groups switched diets for a second session. The men were given pads to wear in their armpits. Havlíček recruited thirty women at fixed intervals of their menstrual cycles to smell the pads and rate the odors for sexual attractiveness and other qualities.

Hands down, the vegetarian diet won. Men's body odor when on the no-meat diet was considered more pleasant, less intense, and more sexually attractive than when on a heavy meat diet. This was true even though the female judges weren't vegetarians themselves.

According to Havlíček, the reason why meat-heavy diets make men stinky has to do with changes in the chemical composition of sweat. Meat contains a high number of aliphatic acids (components of animal fat). Somewhat unappealingly, the strong odor of heavy meat eaters is due to the bacteria that break down these fatty acids. Another possible reason for the stench may be ketosis, the breakdown of body fat, which happens when dieters cut back on carbohydrates and eat a lot of meat. Hormones consumed by or injected into farm animals could also have an effect on body odor, but how much is unknown. Women have more sensitive noses, but it's likely, although untested, that men also have an aversion to meat-heavy odors on women. (Too much red meat reputedly has an adverse effect on oral sex, making sperm, and perhaps vaginal fluids, taste more acidic, while garlic and onions may leave a bitter taste in the mouth.)

The study's result doesn't mean that vegetarians always smell better than meat-eaters. Curries, stinky cheeses, onions, garlic, and dairy products can also make lovers turn up their noses.

Why might you seem less good-looking in a bad-smelling place?

It might surprise you that bad smells are even more powerful than good smells. While a good smell doesn't necessarily mean that a person will be perceived as more physically attractive, a bad smell can make someone appear uglier.

That's the upshot of a study led by experimental psychologist Luisa Demattè at the University of Oxford. She asked women to rate the physical attractiveness of forty men while smelling different odors. In random order, each face was displayed three times with a different olfactory blast: once with an unpleasant odor (rubber or synthetic body odor), once with a good smell (geranium or cologne), and once with clean air (the control). Remarkably, odor really did make a difference in how women perceived the men. When sniffing unpleasant odors, women rated the faces as significantly (about 9 percent) less attractive than when sniffing good odors or neutral clean air. A pleasing aroma didn't make men seem more attractive, but bad smells made them seem less so. (So far, the experiment has not been done with men as judges.)

So what is it about stenches that make people look a little uglier in the eyes of beholders? The Oxford researchers drew on previous odor research to propose that bad smells are distracting and require more brain processing power than good smells. A foul odor triggers the amygdala, the command-and-control center of emotion and fear. When alarmed, a person is more likely to process information, including facial attractiveness, in a negative way. A related theory is that the shift in perception is due to the "halo effect," in which the characteristics of one sensation cross over to another, as when you think of a fruity odor as "sweet."

Although the researchers don't believe the women had any trouble separating sight from smell, it's possible that bad smells somehow leave a residue. Think of it as the difference between standing at the same spot on a sunny beach when the air is clean and fresh versus when it reeks of decaying fish. The bad odor corrupts the way we perceive, or respond to, the beauty of the sea. The same goes when gossipmongers try to detract from a person's image by saying he or she smells bad. Slander like that stinks—and sticks.

CHAPTER 3
A Sound Choice

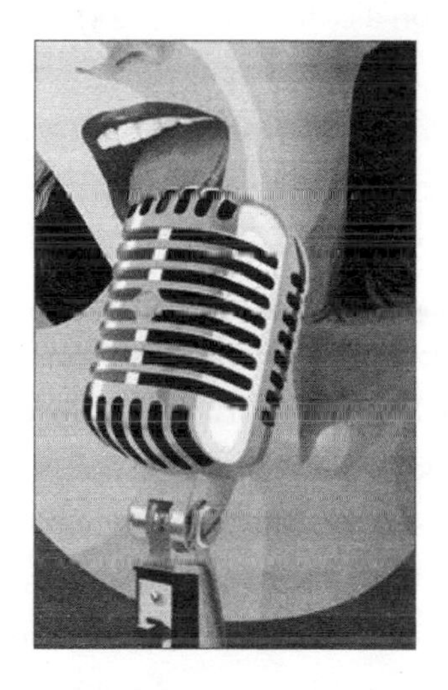

Sing again, with your dear voice revealing
A tone
Of some world far from ours,
Where music and moonlight and feeling
Are one.

—Percy Bysshe Shelley,
from "To Jane: The Keen Stars Were Twinkling"

Why might deep-voiced men have more babies?

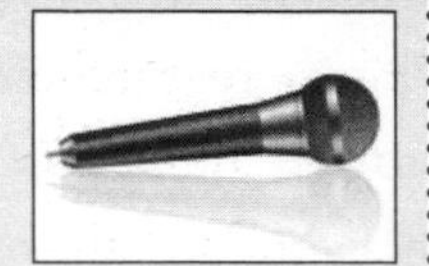

Have you ever been set up with someone who sounds great on paper (or on a computer screen), only to discover when you meet that he has a soft, nasal, gargling voice that drives you to distraction? You know it's low of you, and you can't say anything, but voice can be a deal breaker at the beginning of a relationship. If you're like most women, you prefer deep, rich, husky, resonant voices. A baritone is a boon for a man.

Exploring the connection between vocals and virility, Coren Apicella, an anthropology researcher at Harvard University, and her colleagues David Feinberg and Frank Marlowe interviewed members of the Hadza, a tribe of Tanzanian hunter-gatherers. The Hadza are relatively monogamous and have no access to birth control, which made them ideal subjects. Apicella recorded the voices of Hadza men of all ages and asked them how many kids they'd fathered. When the researchers analyzed the recordings, they saw a clear-cut pattern: deeper-voiced men fathered more children. On average, the men with low-pitched voices had two more children than men with higher-pitched voices.

The connection between men's voices and fertility is the sex hormone testosterone. A deep voice is basically an advertisement of testosterone. At puberty, the hormone makes men's vocal cords thicken, their larynxes descend, and their vocal tracts lengthen. When researchers at the University of Paris at Nanterre took vowel-sound voice samples of young men and played them to female judges, the women accurately guessed that the men with low frequency and small formants (deep and even voices) had higher testosterone levels than their higher-pitched peers. Testosterone also shapes men's musculature and jawline, increases their sex drive, improves sperm quality, and makes for better athletes and hunters. Deep-voiced men may have had higher exposure to testosterone, which would translate into hunkiness, horniness, and, historically, an ability to procure resources, which means . . . well, you get the picture: more babies.

Even more relevant to fertility, researchers have found that around ovulation, when they're most likely to conceive, women have a significantly increased attraction to men's low-pitched voices as well as other high-testosterone cues such as strong masculine features and behaviors. A deep voice may be a cue of social or physical dominance, as found in a study led by biological anthropologist David Puts at Michigan State University. When in competition, men who are perceived as dominant have voices with lower fundamental frequencies and more closely spaced formants (a rich, even timbre). Men perceived as less competitive raise their voice pitch and have looser formants (thinner sound quality) when they feel threatened.

Low-pitched voices are directly associated with high testosterone, although studies find that men with deep voices aren't necessarily larger in size. A great example is my friend Al, a trader on

Wall Street. Al is not a big guy—in fact, he calls himself an "ectomorph," meaning he's fine-boned and has a concave chest, small shoulders, and a fast metabolism. But he looks as tough as can be with a block jaw and a strong brow ridge, and he's a bulldog in the trading pit. He also has a deep, steady, melodic voice that makes the ladies swoon. It's his best quality. And yes, Al's voice drops to a rumbling baritone when he's confronted or stimulated. However, it's not as if every deep-voiced man acts competitive; age, culture, and personal experience temper men but don't raise their voices.

Women's sensitivity to male voices may have been an adaptation that harks back to the time of our remote ancestors, according to Apicella and her colleagues. Historically, females preferred males with testosterone-related traits such as hunting prowess and social and physical dominance—and deep voices are part of the bundle of traits that signal genetic quality. In addition, vocal courtship displays such as sounds and music probably preceded articulate languages and were used to attract mates. Women pay attention to tone and pitch. Genes for deep male voices proliferated as our foremothers chose deep-voiced men and had more babies with them. If it weren't for birth control, it's possible that, everywhere around the world, deep-voiced men would have more children.

Of course, not every man has a deep and soulful voice. But we also fall in love for other reasons: intelligence, creativity, generosity, loyalty, companionship, and so on. Which is why some high-pitched men squeak by.

Why do men prefer high-pitched female voices?

Men prefer women's voices to be like Marilyn Monroe's: somewhat high (but not shrill), full, sweet, sultry, and breathy. That's not to say that the deep, throaty voices of Billie Holiday, Amy Winehouse, and Kathleen Turner don't have their allure. However, in voice attraction studies, men, whether college students or hunter-gatherers, generally go for a dulcet tone with a broad range of up-and-down frequencies.

Exploring the charm of the sweet and high female voice, psychologist David Feinberg at McMaster University in Canada found that men not only prefer high-pitched voices to lower-pitched ones but also choose faces of women with high-pitched voices over those of women with low-pitched voices. Men also rate a raised pitch in a woman's voice as more attractive when she appears interested in the listener than when her attention is directed elsewhere, according to a subsequent study at the University of Aberdeen's Face Research Lab. To men, a feminine high-pitched voice is something of a mating call.

The hormone estrogen is behind the silky elasticity of high-pitched women's voices, and is the reason men are drawn to them. High estrogen levels are linked to female fertility, and women with high levels of the sex hormone have higher conception rates than women with low levels. Estrogen levels dip right before you get your period and plummet more dramatically at menopause. Low estrogen levels result in loss of elasticity in vocal fold tissue, which makes your voice deeper and coarser, sometimes thin and strained and more difficult to control. You might not hear the difference in your voice when you get your period, but the fluctuation can be devastating for professional singers. A chanteuse I met said that, beginning in her early forties, she couldn't hit the high notes the entire week around her period, and has started taking hormones to lessen her symptoms.

A high, sweet voice is but one of many high-estrogen qualities. A study at the State University of New York at Albany found that women whose voices were rated highest in attractiveness were more likely to have pretty faces, more symmetrical bodies, and hourglass figures. Intriguingly, the researchers also found that sexy female voices are a predictor of promiscuity. The more compelling a woman's voice, the more sex partners she was likely to have (in and out of a relationship) and the more likely she was to have seduced men away from their girlfriends. High-pitched women: just call them Sirens.

So smooth, so sweet, so silvery is thy voice. . . .

—Robert Herrick

Test Your Voice

In psychologist David Feinberg's study on the connections between vocal, facial, and body attractiveness, female volunteers were asked to mouth vowel sounds ("eh," "ee," "ah," "oh," "oo") into a microphone connected to a computer. Vocal frequencies were measured using speech analysis software and measured in hertz (Hz). In one experiment, women with low-pitched voices had a mean frequency of 186 Hz, while the sexiest-looking women with the most desirable high-pitched voices had a mean frequency of 236 Hz. (The average male voice is 120 Hz and ranges from 90 to 160 Hz.)

Although the laboratory conditions in the study were carefully controlled, you can do a quick and dirty test of the frequency of your own voice with the same open-source software used by speech analysts. Download Praat software for free online at www.praat.org (available for both Mac and PC). Follow the instructions to record your voice using your computer's microphone. Once you've sampled your voice, go to the menu, click Periodicity, and select To Pitch. Select the new file that pops up, then go to the menu, click Query, and select Get Mean. I scored a middling 201.71 Hz, which makes me feel more like a pigeon than a lovebird.

How does your verbal "body language" reveal your attraction to someone?

When you're trying to communicate your feelings and attitudes, only 7 percent of communication is what you say and 38 percent is how you say it (the rest is body language). Especially in a dating context, it's not the content but other cues—the pitch and pauses, tone, volume, and inflections—that come through loud and clear. In fact, as computer scientist Alex "Sandy" Pentland and his colleagues at MIT Media Lab found, you can be a machine eavesdropping on a couple's first date and—without knowing the meaning of the words or the people involved—predict with shocking accuracy how well the date is going and whether the couple will want to see each other again. All you have to do is listen to all the stuff that's said but not said in a conversation. Think of it as verbal "body language."

Pentland and his colleagues were particularly curious about the parts of speech that are the least in our conscious control. They developed four simple basic "social signaling" measures that reveal how interested people are in each other during a conversation.

Activity level, the first measure, is straightforward. How much are you talking? What's the ratio of your words to his words?

Influence, the second measure, signals engagement. Who's driving the conversation? Is it smooth and equal, or is the other

Master Your Verbal "Body Language"

Among the verbal signals you should be aware of on a date (or anytime) is your pitch (whether your speaking voice is higher or lower than usual), the speed at which words tumble out of your mouth, the ratio of your words to his, the number of pauses in the flow of gab, the tone of your voice, and the amount of mirroring or mimicry going on. Part of the verbal dance involves short interjections that signal listening and agreement. The technology developed by Sandy Pentland and his colleagues at MIT Media Lab counts the number of times people say "uh-huh," "yeah," "yup," "aha," "okay," and "I see" when they're listening to someone speak. In Pentland's study, the more short interjections a man made, the more attractive he was to the woman who was talking. This was especially true when there were a lot of back-and-forth exchanges in the conversation. Mostly unconscious, these "yups" and "yeahs" are ways of mirroring the speaker, showing interest and empathy by encouraging him or her to continue talking. The frequency with which you make interjections suggests something about your interest level (depending upon the pacing; we all know people who can yes-yes-yes along a conversation when they're not interested).

person prodding you along? How long does he pause after you stop talking? Are you cutting off his sentences?

Consistency, the third measure, involves pitch and volume. These factors give away how anxious or uncomfortable you might feel. How loud or soft is your speaking voice throughout the conversation, and how much emphasis are you placing on your words? Are you louder than the other person? Your pitch may also clue people in on whether or not you're extroverted.

Mirroring, the final measure, is how much you and your conversation partner are copying each other's tone and pitch. The more mirroring going on, the more empathetic and in tune you feel. Is he nodding along and saying "uh-huh" as you talk?

According to Pentland's study of sixty 5-minute speed-dating sessions, women's verbal "body language" is more telling than men's when determining whether a couple are romantically attracted to each other and whether they'd like to see each other again. The number one tip-off that a woman was interested in a man for more than friendship was her own speaking rate. Did she talk smoothly and quickly (a good sign), or hesitantly and awkwardly? Women were also more sensitive than men to activity level—how often the two took turns talking—which depends on the balance of empathy and control in the conversation.

Men's romantic interest was predicted most by how much the woman spoke. They were also reportedly more interested in women who varied their tone (up-and-down frequencies). The more engaged the woman was in the conversation, the better the outcome, whereas men's engagement levels were not as significant to the success of the interaction. After all, women are the choosier sex. (There's a caveat to this, as my friend Matt warns. While men on a first date do indeed respond well to chatty women, because volubility seems a good sign of a date's interest and energy level, they don't like too much chatter. Matt once went on a blind date with a woman who talked so much that he started to slouch in his chair and pick at the nap of his corduroy pants. He wanted to talk about himself, too, and prefers women who really seem to be listening.)

By the end of the experiment, the researchers' software was able to predict the outcome of a speed date with an average accuracy of around 70 percent for most couples based on the four speech dynamics alone.

Maybe in the future we'll all wear social signaling monitors that will tell us if a date is into us or not and whether we're sending out the right signals. Until then, try to listen to the verbal "body language" in your conversations. On a subconscious level, you're already doing it. It's called sound judgment.

> *The happiest conversation is that of which nothing is distinctly remembered but a general effect of pleasing impression.*
>
> —Samuel Johnson

Can you tell if a person is gay by the sound of his or her voice?

The "gay male voice" is a stereotype: sometimes higher-pitched, singsong, often laced with lisps and hissed sibilants, and vibrantly expressive. A study led by linguist Janet Pierrehumbert at Northwestern University found that gay men in this subset speak with a more expanded vowel space, especially in words such as *back* and *ask*, than straight men, which may be an effect of slower articulation, emotional expressiveness, and an effort to be precise. Meanwhile, lesbians spoke with more pronounced back vowels (made in the back of the throat) for /u/ and /Ä/ than straight women. (A back vowel /u/ is associated with toughness; think of an exaggerated pronunciation of the /u/ sound in the word *you*, and the /Ä/ in the word *fox*.) Despite popular belief, it's not because gay men's or lesbians' larynxes or vocal tracts are any different from those of straights. Their vocal anatomy is the same as that of straight people of the same gender.

Although you might identify a gay man or lesbian by voice, your detective skills are limited to the obvious. A person who doesn't outwardly identify him- or herself as gay probably wouldn't speak with a "gay voice." The manner of speaking is a learned trait, like an acquired accent, acquired consciously or not from peers, role models, and possibly members of the opposite sex. For some men, such as writer David Sedaris, a lisp and expanded vowels emerge in speech around the same time they start to develop their sexual identity. In his autobiography *Me Talk Pretty One Day*, Sedaris described how the school recruited a speech therapist to eliminate the lisps of fifth-grade boys who, like him, "kept movie star scrapbooks and made their own curtains." (He said it didn't work.)

Gay men and lesbians may effortlessly switch to a "straight" speaking voice. A gay friend said that when he was in high school he was very conscious of his intonation and way of speaking, but now in his thirties he naturally slides back and forth between his "gay voice" and his neutral voice, depending on the context.

Indeed, in one study by linguists at York University and the University of Toronto, straight and gay men read three dramatic and scientific passages and answered open-ended questions in their regular speaking voices. Listening to recordings of the passages, judges thought only half of the gay men actually "sounded gay." Gay judges correctly identified gay men 60 percent of the time, slightly better than a group of mixed male and female listeners. All judges were more accurate when listening to the men answer an open-ended question because "gay voices" were more pronounced. Even so, one of the guys in the study who was pegged as gay was actually straight as an arrow, and many of the straightest-sounding were actually *très* gay.

Does the sound of your name affect how others perceive your looks?

To answer this question, Amy Perfors, a graduate student in cognitive science at MIT, posted frontal shots of twenty-four men and women on HotOrNot.com, a Web site where anonymous viewers rank one another's appearance on a scale of 1 to 10. On each of the photos she added a name such as Jill or Jess and Tom or Ken, and posted the photo multiple times on the Web site under different names, gathering thousands of ratings per face.

In Perfors' preliminary study, the arrangement of vowels in a name did in fact have a subconscious impact on how good-looking its bearer was perceived to be. The difference wasn't huge, but it was statistically significant. Men whose names are pronounced with stressed front vowels, pronounced in the front of the mouth (such as the /a/ in *make* and the /i/ in *beet* or *bit*), were judged as more attractive overall than men with names that were pronounced with stressed back vowels, in the back of the throat (such as the /o/ in *food* and the /u/ in *put* or *tough*). That means the same male faces were rated as slightly hotter when they were given stressed-front-vowel names such as Jake, Ben, Matt, or Rich than when they were given stressed-back-vowel names such as Paul, George, Tom, or Lou. (Say these names out loud to hear the difference between front-of-the-mouth stressed front vowels and back-of-the-throat stressed back vowels.)

The opposite was true for women. When a woman was given a name with a stressed back vowel, such as Laura, Julie, Susan, or Holly, she was considered hotter than when given stressed-front-vowel names such as Liz, Annie, or Jamie. A Zsa Zsa is sexier than a Gigi.

According to Perfors, people might make a subconscious connection between a sound in a language and the meanings associated with that sound. Stressed front vowels may be associated with sensitivity because they're smaller compared to the big, round back vowels that come from down in the gullet. That means we may perceive Jakes and Bens as cuter and gentler than Pauls and Scotts. (To my ear, front-vowel names also sound crisper and more straightforward than back-vowel names.) Meanwhile, female back-vowel names such as Carmen and Susan, at least to me, sound rounder, breathier, and more luxurious than crisp front-vowel names such as Jane, Jen, Tina, and Pat.

Perfors also compared masculine and feminine names. Unsurprisingly, men with masculine-sounding names such as Steve were perceived as more attractive than when given androgynous names such as Jamie and Lee, even though all three have stressed front vowels. Women's looks were also downgraded when given androgynous names.

The popular perception of a name may also rub off on how others perceive your looks. Names such as Brad and Angelina are hot due to celebrity transference. Adolf and Katrina, the führer and the tempest, have ugly connotations. In a name study at Manchester Metropolitan University in England, judges rated women as less good-looking when matched with "unattractive" names such as Tracey than when given "attractive" names such as Danielle. The researchers estimate that nearly 6 percent of the variation in attractiveness ratings of women is attributed to their names. (Men's attractiveness ratings in this study were unaffected.)

"What's in a name?" cries Juliet in Shakespeare's *Romeo and Juliet*. "That which we call a rose / By any other name would smell as sweet." This may be true of roses, but not of people.

CHAPTER 4 *The Racy Parts*

The womb, the teats, nipples, breast-milk, tears, laughter, weeping,
love-looks, love-perturbations and risings,
The voice, articulation, language, whispering, shouting aloud,
Food, drink, pulse, digestion, sweat, sleep, walking, swimming,
Poise on the hips, leaping, reclining, embracing, arm-curving
and tightening,
The continual changes of the flex of the mouth, and around the eyes,
The skin, the sun-burnt shade, freckles, hair,
The curious sympathy one feels, when feeling with the hand
the naked meat of the body,
The circling rivers, the breath, and breathing it in and out,
The beauty of the waist, and thence of the hips,
and thence downward toward the knees . . .

—Walt Whitman, from "I Sing the Body Electric"

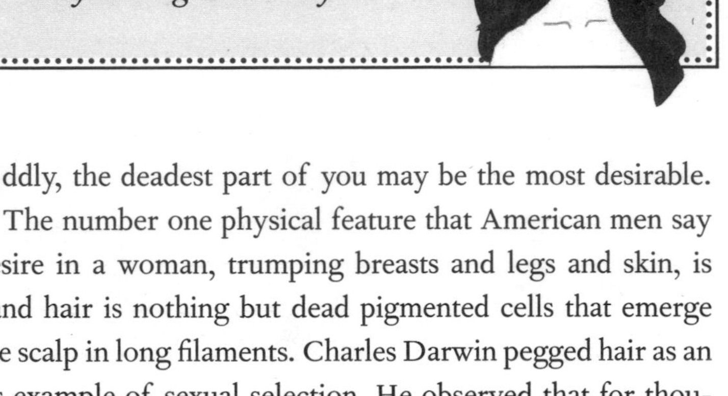

Why is long hair sexy?

Oddly, the deadest part of you may be the most desirable. The number one physical feature that American men say they desire in a woman, trumping breasts and legs and skin, is hair—and hair is nothing but dead pigmented cells that emerge from the scalp in long filaments. Charles Darwin pegged hair as an obvious example of sexual selection. He observed that for thousands of years, in many parts of the world, humans have been able to grow long scalp hair and use it to lure mates. But you've got to wonder—what's so darn sexy about something that doesn't contain a scrap of living tissue?

The answer is that although your hair is dead, it says something about your life. If someone has never met you before and knows nothing about your past, your hair is your best testament. It's an honest record of the food you've eaten, the drugs you've taken, the seasons you've weathered, the stresses and sicknesses you've endured, and the grooming you've given it. Because scalp hair grows on average a half inch every two months, a shimmering mane that swishes below your shoulders advertises at least four years of good health and diet. You couldn't fake that in ancestral times. Our brains subconsciously process the sight of long, vibrant hair as a cue of youth and vitality.

The length and quality of your hair may also reveal whether you've had a child. Hormonal changes in the aftermath of pregnancy may permanently affect a mother's hair, making it shed copiously and turn sparser, darker, coarser, and more brittle—a reason why many new moms get haircuts. In a study of two hundred suburban American women, older ladies and married women with children had much shorter hair of significantly poorer quality, whereas single women without children had longer, more luxuriant tresses. Interestingly, women of all ages who reported health complications that prevented them from having children also had shorter, thinner, weaker hair. Similar results were found in a Hungarian study that compared the effects of various hairstyles (short, medium-length, long, disheveled, knot, bun, etc.) on health and attractiveness. Only the medium and long hairstyles had any positive effect on women's perceived health and attractiveness.

Hair is also important to male attractiveness. While a hairless head is considered sexy on a few exceptional men, women generally don't take a shine to bald guys. In a study at Barry University in Florida, judges of both genders associated balding guys with greater social maturity, wisdom, nurturance, and age—but not good looks. In fact, bald and balding men had negative sex appeal. Likewise, a Korean study with nearly two hundred judges of both genders found that "follicularly challenged" men were perceived as older, less attractive, and less potent than guys with full heads of hair. (I suspect this perception could change as trends inspire cute younger guys to shave their scalps. Many shave their entire head at the first sign of hair loss, believing that chicks prefer an egghead to a comb-over.)

Lastly, there may be a fringe benefit of hair: it broadcasts scent. Hair contains small amounts of fat, and fat absorbs odors easily, which is why your hair retains the scent of shampoo, cigarette smoke—and yourself. Pheromones may emanate from glands that lie at the base of hair follicles, and have an unconscious effect on others (see pages 28–51). Flirtatious women release their scent in the air by caressing, stroking, and twirling their locks, or whipping them around. Your partner will probably want to run his hands through your hair, and you'll probably want to do the same to his. Hair may be long dead, but it does your love life good.

Do gentlemen really prefer blondes?

First, rewind your time-travel clock to the last ice age, some ten thousand to twenty-five thousand years ago, which is when blond hair first spread among the people of northern and eastern Europe. The reason, according to Canadian anthropologist Peter Frost, is that blond hair gave women an edge in competition for a limited resource—men. Few men returned alive from long hunting expeditions. The ones who did were in demand, because women, unable to gather much vegetation in that icy era, depended on men for sustenance. Food was so scarce that polygamy was an option for only the ablest hunters. Forced by circumstance to have only one wife, many men chose a blonde.

According to Frost, ice-age blondes may have been preferred over brunettes because their hair color was rarer and brighter. As we know from walking into any store with an overwhelming surplus, visual merchandising is the key to success. For ancestral Europeans, blond hair was the equivalent of brilliant, shiny packaging. By attracting and retaining men's attention, the dazzling, novel color gave blond women an edge over bland but otherwise equal rivals. Over time, natural blond hair—caused by variants in a gene called MC1R—has reached 40 percent of the Scandinavian population and approximately 2 percent of the world population. And millions more dye to be blond.

Modern men are attracted to blond hair for the same reason as their cavemen counterparts: it's eye-catching. The human eye is attracted to light, bright colors, so blondes stand out more than brunettes and even redheads. As singer Deborah Harry of the group Blondie put it, "As a color it's like walking around with your own spotlight." Men can't help but notice.

Not only is blond hair bright, it's also associated with youth and fertility. Hair color darkens with age. Even women who, as girls, had vivid blond hair often find their locks dulling and darkening over time, beginning in their early twenties. Happily, there's peroxide: 40 percent of the women who dye their hair in the United States choose a shade of blond (not including those who highlight and frost), and in Europe one in every three women lightens her hair. According to a study by Polish psychologists, men clearly prefer blondes when judging the appearance of women older than twenty-five, with the likely reason that blond hair, or even just blond highlights, is "rejuvenating." For some complexions, fair hair minimizes lines and wrinkles and brightens faces, whereas dark hair can have the opposite effect.

Morever, Western media love blondes, which means that blond hair is implicitly associated with glamour and sex appeal. The blonde is perceived as feminine, sexy, carefree, seductive, and capable of having more fun. She's Marilyn Monroe, Barbie, Paris Hilton, Scarlett Johansson, and also Beyoncé and Mary J. Blige. (Women of all races reach for the peroxide bottle today.) A study at Old Dominion University in Virginia found that the hair color of cover models for *Ladies' Home Journal*, *Vogue*, and especially *Playboy* significantly exceeded the proportion of blondes in a random sample of Caucasian American women (which is about 68 percent brunette, 27 percent natural or dyed blond, and 5 percent redhead).

Given all this, do men really prefer blondes? The answer is yes, at least American guys do, according to a study that tracked more than twelve thousand men who used a popular Internet dating service in the United States. The gentlemen indicated a small but statistically significant preference for blondes to women with any other hair color.

According to the rare-color advantage theory, or "novelty effect," hair colors are more desirable when they're uncommon. Blond is usually the standout shade for being unique and the most eye-catching—but not everywhere. In Scandinavia, where blondes are now so abundant that even the brightest manes don't stand out in a crowd, men often say they prefer brunettes. Likewise, when researchers at the University of Washington asked male subjects to choose which woman they'd desire as a partner among selections of brunettes and blondes, the preference for a brunette increased in proportion to the rarity of brunettes in the selection. (However, if a shade is so rare that it's virtually nonexistent, such as blond in Africa and Asia, men may not necessarily prefer it.) Another factor in men's hair color preference is sexual imprinting, the bias for a mate who resembles one's parents. A man with a dark-haired mother might be more likely to choose a brunette for a long-term relationship (see page 19).

Indeed, there's only so much golden tresses can do for you if you accept the wisdom of Anita Loos, whose bestselling book *Gentlemen Prefer Blondes* was the basis of the hit movie starring Marilyn Monroe. Three years after that success, Loos published a sequel. Its title: *But Gentlemen Marry Brunettes.*

Do tall men have prettier girlfriends?

Possibly. Size matters to women—that is, how tall a man stands with his spine erect. Tall men are generally more sexually desirable

Work the "Blonde Effect"

Sig, a guy friend with a history of dating blonde babes—and an annoying habit of telling me I was "cuter" when I dyed my hair blond—admitted that he's simply more attracted to fair-haired women. According to Sig, blondes aren't necessarily prettier, but, as a group, they seem more outgoing, attention-seeking, feminine—and, well, "easier because they're dumber."

The origin of the "dumb blonde" stereotype may be the association of light hair with youth—and therefore ignorance. Interestingly, psychologists at the University of Paris at Nanterre found that among study participants identified as interdependent (i.e., prone to believe stereotypes), those who were exposed to twenty pictures of blonde women before taking a general-knowledge test scored lower than those who were exposed beforehand to pictures of brown-haired people or who weren't primed at all. The psychologists attributed the "dumbing-down" effect to the influence of social context on behavior. Many of us adjust our body language, intellectual performance, and competitiveness to match that of the people around us. If a guy (such as Sig) really believes blondes are dumb, looking at blondes makes him stupider in the same way that priming research subjects with pictures of professors makes them score higher on knowledge tests and priming subjects with pictures of the elderly makes them walk slower, as other studies have shown. Of course, blondes as a group aren't any less intelligent than anyone else. In fact, there are blondes, both authentic and dyed, who make the most of the stereotype—the "blonde effect"—to get whatever they want from their "dumbed-down" beaus, who underestimate them.

than short men. In a study of ten thousand men, the ideal male height was six feet. That's significantly taller than five feet nine inches, the height of the average guy in North America and Europe. Guys who are six feet or taller are more likely to have children than the average-height man, and are also more likely to remarry in middle age and have a second family with a younger wife. In fewer than 1 percent of marriages is a woman taller than her husband. In short, we want men so tall we look up to them even when we're wearing stilettos. (It's mutual: men prefer shorter partners.)

This is lofty news for tall men, who also enjoy other advantages of their height. In some primordial way, height translates into social stature, even in occupations in which you'd think brainpower would prevail. Nearly 60 percent of CEOs of Fortune 500 companies are over six feet tall (and 30 percent are over six feet two inches). It's an unconscious prejudice. Taller people, as a group, are perceived as more intelligent, more dominant, and better leaders. They're also better-paid—and a man's take-home is an important factor in whether a woman will take him home. There's tantalizing evidence that tall men have bigger penises and more sex partners in their lifetimes, and that a woman's preference for height is strongest when she's fertile. All in all, this suggests that tall men are more likely to attract women. Since they're more desirable, beauties are more likely to choose them.

However, pretty women do make exceptions for short men who are outstanding in other ways. In a study of dating trade-offs based on data from twenty-two thousand singles on an online dating service, women were willing to, well, *overlook* men's height limitations provided that their status and resources offset them. Based on dating patterns, a successful five-foot-five guy needs to make $237,500 a year to be as desirable as a six-foot guy who pulls in an average annual salary of $62,500, all else being equal. When wealth compensates, short guys aren't shortchanged.

Why are high heels sexy?

"I don't know who invented high heels, but all women owe him a lot." That's what Marilyn Monroe said about her stilettos, and she has a point. Heels make you statuesque. Your feet look smaller and your gait is more refined. Your calves and shins are tensed and elongated. Your posture is bolt upright.

Anatomically speaking, in heels you're doing what chimps do

when they're in heat: standing on tiptoe, arching the back, and sticking out the butt. The movement of your lower limbs becomes more sensual. It's hard for others not to notice the sway of your hips, the thrust of your breasts, the incline of your pelvis, the strut of your stuff.

High-heeled shoes adjust women's body proportions to come closer to perceived ideals (in Western countries, at least). Researchers at the University of Wroclaw in Poland asked more than two hundred men and women to rate the attractiveness of diagrams of seven men and seven women with varying leg lengths. Both sexes agreed that a leg length that is 5 percent longer than the norm for a person's height is ideal. This means that if the average leg length of a five-foot-five woman is 30 inches, as measured from the sole to the crease where the thigh meets the pelvis, a gal this height could make her legs look 5 percent longer by wearing 1.5-inch heels. In the study, legs that were 10 percent longer than average were also considered sexy, but legs 15 percent longer were not. Generally speaking, if you're five-foot-four to five-foot-eight, heels up to 3–3.5 inches will flatter your proportions, but higher heels may be a stretch.

Also focusing on proportions, a study at University College in London found that the ideal female figure had legs exactly 1.4 times the length of the upper body, which happens to be the leg-to-torso ratio of Nicole Kidman, Naomi Campbell, and most other supermodels. When five-foot-eleven Kidman and five-foot-seven Tom Cruise divorced, she said with palpable relief, "Now I can wear heels." Not that she needs them.

One reason why long legs are so sexy is that they're a cue of health and stable development. When disease, malnutrition, or genetic mutation disrupts growth, the legs are usually stunted in proportion to the trunk. A longer leg-to-torso ratio is linked with better health, including a lower risk of heart disease, diabetes, low blood pressure, and even cancer.

Going out on another limb, evolutionary biology's "handicap hypothesis" might also explain men's fetish for women in heels. It goes like this: In nature, any display of excess is a handicap. Bright, heavy tail feathers are a handicap to peacocks; they tax the bird's resources and attract predators. Similarly, high heels are a handicap to women; they're difficult to walk in and require good joints and coordination. Heavy tail feathers and tricky high heels are ways of flaunting fitness in that one must be in good physical shape to "afford" them—and fitness is very sexy. Seen this way, high-heeled shoes are something of a Darwinian challenge. Stay on your toes, literally—to attract a mate and pass on your genes.

What does a "wiggle" in your walk reveal?

The more wiggle in your walk, the tinier your waist is in proportion to your hips—a telltale sign of youth and fertility. It's what James Taylor was getting at when he crooned, "There's something in the way she moves." That something is *sex appeal*. Taking this to the logical next step, you probably think your walk is most provocative during the fertile days of your menstrual cycle, right? Actually, no. To the contrary, it turns out that's when a woman's gait is most restrained. When you're most likely to get pregnant, your hips might not roll as much and your knees are closer together.

These results shocked the authors of the study, evolutionary psychologist Meghan Provost and her colleagues at Queen's University in Canada. In fact, they were so surprised that they confirmed them twice over. Provost had recruited female volunteers and recorded them, at different stages of their menstrual cycles, walking in suits with light markers along the joints and limbs. She showed male raters animations of the walking figures (light points only, no actual bodies, thanks to the magic of computer modeling software) and asked them to judge the sexiness of their strides. Each time, the men judged the least fertile women as having the sexiest walks. When women were fertile they walked in a more discreet close-kneed manner, with less swivel and saunter than usual. Evidently, hormones related to ovulation have an effect on a woman's gait, probably by acting on joints and ligaments—but not in the expected way.

Provost speculates that the reason women walk less provocatively when they're most fertile is an unconscious attempt to avoid excessive male attention. While subtle cues of ovulation may be perceived—in estrogen-plumped facial features, a higher sex drive, and even your smell—these sexy signals would be picked up only by men who are in close range. As a protection against assault, our foremothers may have adopted a less attention-getting walk when fertile. In an ancestral environment, it would have been in a woman's best interest to have babies with chosen partners rather than a stranger or an undesirable male who caught a glimpse of her from afar and forced himself on her.

Then again, it's all conjecture. A hip-swinging saunter is perceived by Western cultures as sexy, but not all cultures may agree. Low waist-to-hip ratios that give women a wigglier walk are desirable in the industrialized West but might not be in parts of the world where thicker waists are considered more attractive. In other walks of life, different strides could be sexier.

Make Your Moves

Models, dancers, and actors are trained to pay attention to the way they move, and so should you if you're curious about what a gait can do for you. In a joint New York University (NYU) and Texas A&M study involving more than seven hundred judges, the sexiest female walks involved a sway—that is, lots of side-to-side motion in the hips. You can improve your hip-rolling wiggle by turning your feet out in a V shape with your heels closer and your knees slightly farther apart. As you walk, roll your foot from heel to ball, and move the bent leg toward the weight-bearing leg. Keep your upper body steady. Develop a rhythm so it feels as if your hips are moving in a circular motion. High-heeled shoes may make the sway more pronounced.

Hormones not only affect your walk but also influence your attraction to men's walks. In a subsequent study on male gait, Provost found that women, regardless of ethnicity, were attracted to male walks defined as "more forward-and-backward motion of the shoulders relative to the hips." (Meanwhile, gay men can be identified with about 60 percent accuracy from their wigglier walk, according to an NYU study, because some have lower waist-to-hip ratios—larger hips compared to waist—than the average straight guy.) Interestingly, women in their fertile phase more strongly preferred very masculine shoulder-rolling gaits. The more inclined they were to have a fling or one-night stand, the more attractive they found it—so watch yourself when you swoon over a swagger.

Why are curves sexy?

From a hundred feet away, a man can't see your beautiful eyes or your luscious lips. He can't hear your witty jokes or touch your dewy skin. Nor can he hear your sultry voice or smell your sweet scent. He might never get to know you, but by merely glancing at your figure he'll glean a lot about your age, health, and reproductive potential (which is sufficient for some Casanovas). That's because he can instantly assess your waist-to-hip ratio (WHR).

A woman's waist-to-hip ratio is one of the most important cues in sexual attraction. The smaller your waist is in proportion to your hips, the curvier you appear. The "golden ratio" is said to be around .7—that is, a waist that is seven-tenths the width of the hips, regardless of weight. That's the approximate WHR of female flint figurines unearthed from prehistoric sites, statues of the fertility goddess Venus, the wasp-waisted dancing girls in ancient Hindu paintings, the corseted ladies of old Europe, and beauties such as Twiggy, Kate Moss, and Marilyn Monroe. (Yes, slender and buxom women may have the same WHR.)

A WHR of .7 is considered the most desirable by men, as found in studies led by evolutionary psychologist Devendra Singh at the University of Texas at Austin. Singh and his colleagues showed men ages twenty-five to eighty-five line drawings of twelve female figures of varying WHRs and body weights. They asked the guys to rank the drawings in order of attractiveness. From the cocky young upstart to the doddering octogenarian, Western men pretty much unanimously agreed on what female body types are hottest and healthiest. The highest ranking went to the normal-weight figures with a relatively low WHR, around .7, which is rather curvy. In Singh's subsequent study that included breast size, young men overwhelmingly said they'd prefer to date big-breasted women with small waists. Although there is a preference for big hips throughout the world, some studies have found that men from poor, rural backgrounds prefer women with a WHR around .8, representing a somewhat less curvy body type with more cushion around the midsection. (See pages 106–107 on economic influences.)

From an evolutionary view, curves are desirable because they're blatant cues of youth, health, and fertility. The hourglass figure is unique to a woman in her peak reproductive years, and often vanishes after her first baby. Wide hips are crucial because babies' big heads require a lot of pelvic space. From kimonos to corsets, waist-cinching fashions are sexy because they make the hips and rear appear wide in contrast. Our foremothers stored fat on their hips, butts, thighs, and breasts to sustain the caloric requirements of pregnancy and nursing, and women usually tap into these fat reserves only at these times, even if they are malnourished.

Harvard biological anthropologist Grażyna Jasieńska found direct evidence that the hourglass figure is related to fertility. Measuring 120 women of childbearing age, she discovered that women with low WHR ratios and relatively large breasts had 37 percent higher levels of estradiol (an estrogen hormone) during the fertile part of their cycles and 26 percent higher levels during the remainder of the cycle than did women who had higher WHR ratios (not as curvy). Estrogen increases fertility and influences where and how fat is deposited on your body. Thanks to estrogen, fat doesn't accumulate on the waistlines of young women as it does on those of men and older women. In Jasieńska's study, women with curvy figures were roughly *three times* as likely to conceive as women of the same age with more tubular figures—a shocking difference.

Intriguingly, there's also evidence that men evolved to prefer women with a low WHR because curves mean smarter kids. William Hassek, an epidemiologist at the University of Pittsburgh, and Steven Gaulin, an anthropologist at the University of California at Santa Barbara, have drawn attention to the fact that fat stored in the hips and thighs is a different and much healthier fat than the pudge around the waist. Hip and thigh fat ("gluteo-femoral fat") is rich with long-chain polyunsaturated fatty acids and omega-3 fatty acids (including DHA) that are essential to fetal brain development. When women are pregnant or breast-feeding, these "smart fats" feed and nourish the baby. They constitute 20 percent of the dry weight of the human brain, and a baby's IQ is known to increase .13 points for every 100-milligram increase in daily intake of DHA alone. After trawling through data from 1,933 mother-child pairs and controlling for factors such as age at first birth, education, father's education, income, and race, the researchers found that children whose mothers had a low WHR really are smarter. Mothers' WHR alone accounted for nearly a 3 percent variance in children's scores on four cognitive tests.

If nothing else, a WHR around .7 is sexy because a ratio above .9 is a sign of poor health or infertility. A high WHR may be a sign of menopause or reproductive disorders, with decreased estrogen levels or elevated testosterone levels to blame. It may also be indicative of problems such as cardiovascular disorders, diabetes, or gallbladder disease. Then again, you could have a high WHR and be youthful and healthy . . . and pregnant.

Measure Your WHR

Stand, breathe normally, and wrap a cloth measuring tape around the narrowest part of your waist, which is a bit above your navel. (Try to resist sucking in your breath.) Write down the number. Then get your hip measurement by placing the tape at the widest part of your hips and buttocks. Write down that number, too. Then calculate your WHR by dividing the waist measurement by the hip measurement.

WHR = waist measurement / hip measurement

Don't fret if you don't have Marilyn Monroe's perfect .7 figure. You can dress to make it seem as if you do. If your hips and waist are about the same, wear shirts, blazers, and dresses that are tailored on top but that flare out at the waist, to give the illusion of wider hips. If your waist is larger than your hips, reduce your waistline by wearing vertical stripes, blazers, and V-neck shirts, which accentuate the top and deemphasize your bottom half.

Why do men love big breasts?

Mammary glands turn off all male mammals except men. That's because big teats signal temporary infertility in all female mammals except women. Out they pop on a pregnant or lactating dog or cat, a sign of sexual unavailability, and in they go, mostly, after the pups and kitties are weaned. But women's breasts poke out at puberty and never go away. Guys find this totally titillating.

Nobody has a definitive answer as to why human breasts are so sexy and get so big, but all theories have something to do with fertility. Evolutionary psychologists suggest that cleavage serves as a sort of proxy for the swollen rumps that other female primates get when they're in heat. Freudian psychologists offer theories about men's Oedipus complex: they're always looking for a mother figure (literally). Anthropologists believe that women developed larger, permanent breasts as our species adapted to a harsher environment and became bigger-brained and bipedal. By storing fat reserves in their chests (and thighs and butts) year-round, even when not nursing, our foremothers survived the elements and the rigors of pregnancy, birth, and child rearing.

Large breasts may be a sign of increased fertility, which could help explain why so many men think bigger boobs are better. The fat that accumulates in your chest (as well as butt, thighs, and hips) does so under the influence of the hormone estrogen, which also affects your ability to conceive. As mentioned on page 95, a study by Harvard epidemiologist Grażyna Jasieńska found that full-figured women are roughly three times as likely to get pregnant as women with other body types. (To qualify as big-breasted in the study, the circumference of your torso around your breasts would have to be at least 20 percent larger than it is under your breasts.)

Breasts are also an advertisement of age, health, and good genes, which is why anthropologists think they're crucial to sexual selection even in cultures that don't eroticize the chest any more than the face. The larger the boobs, the more they reveal about you because women's chests have an unfortunate tendency to sag. In pre–plastic surgery days, only very young women had breasts that were both big and firm.

The symmetry of a woman's breasts also reveals something about her health. Because breasts grow so rapidly during puberty, they're especially sensitive to hormonal disruptions that could make one breast grow much larger than the other and also compromise fertility. Women with asymmetrical breasts have fewer babies on average than do women with symmetrical breasts. Worse, the odds of developing breast cancer increase with breast volume asymmetry. For every 100-milliliter difference in size between breasts, women's risk increases by 50 percent. (Don't panic! Most women have minor breast asymmetries. On average, the left breast is 4 percent larger than the right.)

Big chests are part of a suite of sexy feminine features associated with health and higher estrogen levels, including a low WHR, rounder facial features with a smaller chin, and a higher-pitched voice. Environmental factors also influence breast size. In China, the average cup size has increased by nearly an inch in a decade, with more Chinese women wearing B and C cups than ever before, while the average in America went from 34B to 36C. Fattier diets are the culprit. Yet when women say they want to lose weight, they rarely mean the fat stored in their chests.

Why do women feel pressured to be superskinny?

Wallis Simpson, the Duchess of Windsor, was at least half wrong when she said a woman can't be too rich or too thin. No matter how loaded you are, it's better to be a healthy average weight with a body mass index (BMI) between 18.25 and 24.9. Men prefer it, at least theoretically.

When evolutionary psychologist Devendra Singh asked men of various races and ethnicities—including African American, white American, Indonesian, and Indian—to pick the most attractive female body type out of a series of line drawings, their universal first choice was the curvy medium-weight figure. This makes sense, Singh concluded, because overweight and underweight women are more likely to develop health problems. If you're X-ray thin—or even just 10–15 percent below the normal body mass for your height—you may be temporarily infertile. Your body prevents you from getting pregnant when you're physically unable to support a child growing inside you. Your womb shrinks,

Measure Your BMI

Your body mass index determines if you're underweight, normal, or overweight for your height. BMI is an international standard, and the official metric is weight in kilograms divided by your height in meters squared:

$$BMI = \text{Weight in kilograms} / (\text{Height in meters})^2$$

To get your BMI using U.S. standards, divide your weight in pounds by your height in inches and multiply by 703:

$$BMI = (\text{Weight in pounds}) / (\text{Height in inches})^2 \times 703$$

Or use the online tool at the Centers for Disease Control and Prevention: www.cdc.gov/nccdphp/dnpa/bmi.

BMI	
< 18.5	Underweight
18.5–24.9	Normal
25.0–29.9	Overweight
30.0–34.9	Obese
35.0–39.9	Very obese
>40	Extremely obese

your periods dry up, and your sex drive slows down. In extreme cases it stops. How sexy is that?

So, if not to attract men, why do the Victoria Beckhams and Kate Mosses—and perhaps you, too—strive to be superthin? It's complicated. Blame the media and all that's related to them: the fashion industry, the diet industry, and the Hollywood ideal to which so many women aspire. Fault the economic and social and psychological stresses that trigger eating disorders. Point to our desire to compete with other women, using anorexic models as a sort of measuring stick. Some go so far as to suggest that a shift in focus to career instead of marriage and motherhood has something to do with it. Psychologist Nancy Etcoff, in her book *Survival of the Prettiest*, described the provocative theory that extreme dieting suspends fertility by slowing down the aging of the ovaries. This could mean women are acting out an unconscious strategy to delay having children until their career or romantic ambitions have been met.

Most of these reasons shrink down to media, culture, and status. You can't deny that socioeconomic factors play a role in what is considered attractive. Remember, suntanned skin was considered nasty when it was associated with laborers yet sexy when associated with the idle rich. The same goes for skinniness, which is attractive because it's now associated with social and economic advantage, not poverty. Cheap fast food makes people flabby, and the upper crust differentiates itself by skipping dessert. It follows that men who care about social status don't mind having stick-figure girlfriends. (Interestingly, studies have found that American black men, as a group, prefer heavier women. This may help explain why fewer black women have anorexia.)

The skinny here is that culture can trump biology. Sometimes the perceived ideal body type is not an average healthy one—or, ironically, the one with the most sex appeal.

Why do men feel pressured to be buff?

Men's bodies are also under scrutiny, although the pressure for the average straight guy to be in shape isn't as intense as it is for women. Frankly, women aren't as likely to reject a man just because he doesn't meet a physical ideal (but looks count, of course, sometimes more than women are willing to admit). Ladies prefer guys with a V shape—narrow waist and broad shoulders—which loosely translates into a waist-to-hip ratio (WHR) around .90 and no higher than 1.00, and a waist-to-shoulder ratio (WSR) of .75 or lower. Young men have an ideal muscle mass that is about 43 percent of their body mass.

Given that women are more lenient than men about their partners' looks, why do so many men want to be so muscular? The reason, according to a study by psychologists at UCLA, is that men increasingly face within-gender "prestige competitions." Just as the media perpetuate an image of the female body that is excessively skinny, so it pushes a male ideal that is excessively ripped. Undressed guys featured in men's glossies such as *Men's Health* are significantly more muscular than the hunks in, say, *Cosmopolitan*. Men in Western countries have begun to feel as if the masculine norm is the Abercrombie & Fitch guy with six-pack abs, bulging biceps, and deltoids on steroids.

Moreover, they think women like it. Just as many women think men prefer superskinny women, men think women prefer ultra-buff men. For long-term relationships in particular, these assumptions are wrong. According to a study by biopsychiatrists at Harvard Medical School, Western men believe women desire a male body that has about 20–30 pounds more muscle mass than their own and the male average. However, when undergraduate women were asked which body type they preferred from a lineup of male figures with various physiques, they picked one that was toned but without all the bulk and brawn. While men don't desire extreme female thinness because it undermines their subconscious preference for fertility, women don't desire *extreme* muscle because it undermines their preference for lower-testosterone qualities such as loyalty and paternal interests. If a muscle-bound guy is desired, it's more likely for a short-term relationship. According to studies at UCLA, 61 percent of women report that their flings were brawnier than their boyfriends.

Interestingly, Asian men in Taiwan were more comfortable with their bodies than their Western peers, less likely to desire more muscle, and less exposed to images of buff males in the media. This leads researchers to wonder if the hypermasculation of the Western male is a backlash against feminism. Bulking up is the only thing guys can do that most women cannot.

Why do hungry men prefer heavier women?

Take a look at *Playboy* magazine's Playmates of the Year for a sign of how the economy is doing. That's essentially what psychologists Terry Pettijohn II and Brian Jungeberg did, by comparing the sizes of centerfolds' chests, limbs, and facial features through economic booms and busts. They found that in years when the consumer price index, unemployment, homicide rate, and other measures indicated economic hard times, Playmates were slightly heavier, older, and taller, and had larger waists and chins and smaller eyes. (In 1993, a disastrous year, Anna Nicole Smith graced the centerfold.) When social and economic conditions improved, Playmates were lighter, smaller, shorter, bigger-eyed, and smaller-waisted.

Hard times also seem to subtly shift men's preferences in women's faces. Evolutionary psychologist Anthony Little at the Face Research Lab at the University of Liverpool asked participants to imagine themselves in a harsh condition in which they are uneducated, hungry, jobless, and living in a violent place without social support. Following that, they were asked to rate the attractiveness of opposite-sex faces and determine which they prefer under what conditions.

In harsh conditions, for long-term relationships, men preferred women's faces that were tougher-looking and more masculine, and for short-term relationships they preferred more delicate-looking, feminine faces with bigger eyes, button noses, smaller chins, and rounder babyish cheeks. When the going gets tough, men seem more comfortable settling down with more mature-looking women. They're perceived as stable, strong, solid, independent, and higher-status—perhaps better partners if you're a guy who wants someone to take care of you. When conditions are less threatening and resources abundant, men's overall preference swings back to women with feminine features such as big eyes and lighter, smaller, weaker bodies. (Women prefer gentler-looking, less-masculine men for a long-term relationship in harsh conditions; they're perceived as more loyal and supportive. For short-term relationships, women prefer very masculine-looking men. These explicit preferences—gentlemen for marriage and studs for one-night stands—are generally true in the best of times and the worst of times.)

Given the results of Little's study, it's unsurprising that in some parts of the developing world where resources are scarce, men think thicker-waisted women are sexiest. While we in the West pick at celery and salads, our counterparts in Mauritania eat butter and cream, force-fed by relatives who think more body fat makes young women marriageable. Male attraction to a thicker waist is a reaction to tough economic conditions, according to evolutionary psychologists Viren Swami and Martin Tovée at the Universities of Liverpool and Newcastle. In a study of female body type preferences, they asked European and Asian men of varying economic backgrounds to rate the attractiveness of women with known body mass indexes. The guys from industrialized countries, regardless of race, preferred normal-weight women. Meanwhile, men from poor rural areas chose ladies with heavier body types. (In some developing countries, such as Kenya, studies have found that men desire women with a low waist-to-hip ratio albeit with a heavier body mass.)

Weirder yet, men's aesthetic tastes may change depending on how hungry and secure they feel at any given moment. Testing this theory, Swami and Tovée stopped sixty undergraduate men outside a dining hall at an English university and asked them how hungry they felt. Approximately half the men said they were starving, while the other half was leaving and their stomachs were full. The researchers showed them photos of fifty women with various body types and asked them to rate the women's figures (bodies only, no faces). Although all guys preferred body types that fell within an average weight range, the ravenous ones preferred heavier figures. (The peak BMI preference of guys with empty stomachs was 22.97 versus 20.72 for the guys with full stomachs.) When asked to rate the attractiveness of obese women, hungry men gave them higher ratings than did guys with full bellies.

According to Swami and Tovée, men's ideals unconsciously shift according to what is practical in different environments. For all of us, our drives and decisions also depend on blood sugar and hormone levels, which reflect how recently we've eaten or how calm and secure we feel. In a time of scarcity, men's biological instinct is to be drawn to women who appear fertile and able to cope with manual labor, pregnancy, and childbirth. Body fat may be psychologically soothing in that it represents abundance and capability.

So, is your beauty as others see it influenced by per capita income, terrorism, the Dow Jones Industrial Average, the crime rate, where you live, and whether breakfast was skipped? Perhaps in subtle ways, although no studies have been done on how economic conditions affect real-life mate choices. But maybe it's worth picking up a copy of the *Wall Street Journal*, or *Playboy*, to see where you stand.

Why do so many men wish they had bigger penises?

The woman's vagina in fact is so cleverly constructed that it will accommodate itself to each and every penis; it will go out to meet a short one, retire before a long one, dilate for a fat one, and constrict for a thin one. Nature has taken account of every variety of penis and so there will be no need to seek a scabbard the same size as your knife.

Thus concluded Reinier de Graaf, the seventeenth-century Dutch anatomist, on whether penis size matters when it comes to sex. The subject was put to bed, but not to rest, and over the centuries it has shown remarkable stamina. Limp male egos have spawned a huge industry in penis enlargement. By pill, by pump, or by knife, many men aim to enlarge their penises just as women enlarge their breasts. It's to satisfy their partners, or so they say.

Worldwide, the average erect penis is 5.3 inches long, with 68 percent of men measuring in the normal range between 4.6 and 6.0 inches. The standard girth is around 4.8 inches. By comparison, a woman's vagina is only about 3–4 inches long on average, although it expands and accommodates when she's aroused. (Women who have given birth vaginally have a larger vaginal opening, but toned pelvic floor muscles may keep the insides strong.) Moreover, women have nerve endings only in the first 2.5 inches of their vaginas.

For these anatomical reasons, women are basically satisfied with the shapes and sizes of the various penises they encounter. If men feel underendowed, they should check out the Sex and Body Image Survey, a study undertaken by sociologists at California State and UCLA of more than fifty-two thousand worldwide visitors to the MSNBC.com and Elle.com Web sites. Only 6 percent of women considered their partner's penis too small, and only 14 percent of women were "size queens" who wished their partners had larger penises. Pose the penis size question to a group of girlfriends over drinks, as I did, and chances are the consensus will be no, it doesn't matter. One friend, an opinionated banker, slammed her beer down, gave a brief laugh, and said, "Skill can compensate for a small penis, but size can't compensate for lack of skill." Even so, many men with average-sized penises perceive them as a serious shortcoming. Nearly half the men in the survey—45 percent—said they'd like their own to be larger, even though only 12 percent considered themselves small.

Why do so many men wish they had bigger genitals when so few women care? One explanation is that men may unconsciously desire to increase their fertility. Some biologists suggest that larger penises evolved in humans (other primates have smaller ones) to increase men's chances of impregnating women. A long penis extends deeper and squirts semen directly into the cervix, and may be shaped to scoop out rivals' sperm (see page 302). A wide penis could provide a better "plug" that prevents semen seepage. A thick dick that rubs against the clitoris may also trigger a woman's orgasm, which could increase a man's chances of successful conception.

But there's another, more covert explanation as to why so many men wish for a magic wand to make their penises grow, as suggested by the sociologists who analyzed the survey results. Penis size is universally associated with masculinity. As long as the penis is a symbol of power, it has power over men. Just as the media emphasize the allure of big breasts and skinny bodies for women, so it pushes plump pecs and penises at men. Ironically, women actually aren't as attracted to either as men think (see page 104).

But here men might not have women first in mind. According to evolutionary biologist Jared Diamond in his book *The Third Chimpanzee*, the penis has as much to do with status as it does sex. The bigger it is, the mightier and more manly the guy. As a symbol of virility, its historical role is that of a threat or status display tar-

Don't Look at His Hands

In case you *do* care, there's only one anatomical feature that (very) generally correlates with penis size—and it's not the hand. In a study of thirty-three hundred Italian men, researchers found that only *height* was correlated with larger penises, meaning the taller the man, the larger the penis. Height as a predictor of penis size was also supported in the results of the more than twenty-five thousand men profiled in the MSNBC/Elle.com survey: compared to men of average height, significantly more guys who were shorter than five feet eight inches reported small penises and significantly more men who were taller than six feet reported large penises. And here's another myth debunked: race is not a reliable indicator of penis size. Multiple studies have not found any consistent correlation between race, ethnicity, and penis size. There is much more variation *within* racial populations than *between* them, according to a World Health Organization study. If you want to maximize your chances of finding a man with a large penis, your best bet is to stop looking at hands and skin color and start looking up.

geted at other men, not unlike how a lion's mane threatens other lions. This helps explain why words such as *manhood* become synonymous with *penis,* and "big swinging dick" becomes an honorific.

In short, there's no doubt that men are under pressure to measure up in boardrooms and locker rooms. But at least they might get a break in the bedroom.

Are men with large genitals more likely to cheat?

No, there is no hard proof that men with large genitals (testes or penis) are bigger cheaters than their smaller-endowed peers. This misconception may come from the theory of sperm competition, which is that highly promiscuous species have large testes. Chimpanzees, for instance, are big swingers and have melon-sized balls. The larger the testes, the more sperm they hold. Promiscuous animals need to pump out a lot of sperm because females may mate several times per day with different males, and the male who ejaculates the most has a better chance at winning the fatherhood prize. However, this doesn't apply to testis size *within* a species. While men with large testes do indeed ejaculate more sperm than guys with small testes and therefore could theoretically sire more children, studies have determined that they are *not* more promiscuous.

The same goes for penis size. Men with large penises are not more inclined to be unfaithful than the average Joe (or Dick). In this way for sure, size doesn't matter.

What's the purpose of pubic hair?

When evolution stripped our bodies of hair, probably to cool us down and reduce parasite infestation, it neglected the hair on our heads, armpits, and pubes, and on men's faces. The head hair makes simple sense: it protects us from hot and cold, and displays our nutritional and developmental status. Men's beards are also easy to explain: they protect against cold and are a display of masculinity. But the purpose of pubic hair is a little more covert.

One possible purpose of pubic hair is that it protects the genitals from debris, much in the way that eyelashes do for the eyes. There's also speculation that the hair "down there" has played a role in sexual selection. For one, it's a blatant cue of fertility—only a person who has reached sexual maturity grows pubic hair. Pubes may also disseminate pheromones. Sweat glands lie at the base of each hair follicle, and these glands excrete compounds that may contain odors that are unique to you. Hair contains some fat, and fat absorbs body odors, which is why pubic hair retains personal odor so well. Pubic hair follicles also act like antennae, broadcasting a person's scent to anyone whose nose happens to be in close proximity (see pages 28–49).

Strip Tease

For some women, their "pubic hairstyle" is considered just as important as their hairdo. In a survey of nearly fourteen hundred women conducted by the market research firm Harris Interactive, 25 percent indicated that they closely trim their pubic hair; 23 percent say they shave or wax part of their pubic hair, leaving a "landing strip"; and 9 percent remove it altogether (the "Brazilian" bikini wax). Together, this means that more than half of women consider their privates grounds for a beauty ritual. Meanwhile, an Internet survey of approximately seventy-two thousand men at AskMen.com revealed that 39 percent of guys prefer partners who are clean-shaven down there, 38 percent prefer trimmed pubic hair, and only 23 percent go for *au naturel,* or have no preference. It's possible that good grooming displays healthy habits, and sparse pubes suggest youth. The feminist view of pubic hair waxing, especially the bare-as-a-baby Brazilian treatment, is that it smacks of infantilizing women.

PART II
BEHAVIOR

CHAPTER 5
His-and-Hers Hormones

Science—

beyond pheromones, hormones, aesthetics of bone,
every time I make love for love's sake alone,

I betray you.

—Katherine Larson, from "Love at Thirty-two Degrees"

Why are there days when men seem especially drawn to you?

Want to know the best time to go on a big date? Take a cue from women who rely on sex appeal to make a living. Among strippers employed by Albuquerque strip clubs, those who lap-danced in the fertile phase of their menstrual cycles earned an average of $70 in hourly tips, compared to $35 when they had their periods. During the rest of the month the dancers averaged $50 an hour in tips. Those who took birth control pills made significantly less money throughout their cycles.

The likely reason for the ups and downs in lap dancers' income, according to evolutionary psychologist Geoffrey Miller, the study's lead author, is that hormones affect the way women look and feel and act—and we look, feel, and act sexiest when we're most likely to get pregnant, in midcycle, in the days leading up to ovulation.

The next time you're in the middle of your menstrual cycle, look at yourself in the mirror. You might come to the same conclusion as did psychologist Miriam Law Smith and her colleagues at the University of St. Andrews in Scotland. Studying the influence of hormones on facial appearance, the researchers took photos of fifty-nine young women every week for six weeks (bare faces only, not hair, body, or clothes) and gave them tests to pinpoint the timing of ovulation. Then they recruited nearly thirty volunteers to rate the faces. The judges, both male and female, rated the faces of women who were in the fertile phase of their cycles as significantly more attractive. Of two facial composites, the "fertile face" made up of ten women's faces when ovulating was also considered much prettier than the composite made of ten women's faces when not fertile.

It sounds like science fiction, but the hormone estrogen, which surges right before ovulation, has soft-tissue-building properties that subtly "rebuild" your face. The estrogen surge makes lips fuller and skin plumper and smoother. There's also evidence that higher estrogen levels produce healthier lip color, larger pupils, and a more even complexion. In the days leading up to ovulation, the soft tissue of your ears, fingers, and breasts also becomes more symmetrical than during the rest of your cycle. If you let a guy get close enough to smell your underarms or lick your genitals, or if you French-kiss him, he may like your smell and taste more than at any other time of the month.

Your outfits and accessories may be other telltale signs of ovulation. Evolutionary psychologist Martie Haselton and her colleagues at UCLA asked nearly forty volunteers to look at full-length photos of thirty women taken at different times of their individual cycles and judge if the women were trying to look more attractive in one photo than the others. The women featured in the photos did not know the purpose of the study, so they had no reason to dress differently, and their heads were blocked out to prevent raters from mistakenly rating their facial attractiveness. Better than by chance—nearly 60 percent of the time—the raters were able to guess if a woman was in her fertile phase based on her clothes alone. The tip-off was that the women were more likely to wear sexy clothes and expose more skin. They'd often wear outfits that were similar to what they were wearing in a low-fertility phase, but swap pants for a skirt or add feminine touches such as lace-trimmed tops, a frilly neck scarf, or sparkly jewelry. Similar results were found in a German study of club-going women: those who were ovulating exposed more skin and wore tighter, shorter clothes than women who weren't ovulating.

You may also become more horny and flirtatious when you're most fertile. Susan Bullivant, a researcher in biological psychology at the University of Chicago, led a study that found that a woman's sexual desire and erotic fantasies intensify in the few days before she ovulates. Bullivant and her colleagues concluded that the revved-up sex drive in this phase is due to already-high levels of estrogen coinciding with a brief spike in the hormone testosterone. Testosterone amps up the libido.

You might be aware of these monthly fluctuations, these twinges and whims, but only if you're really self-vigilant—and not taking oral contraceptives. The hormones that give you the glow are linked to ovulation, and birth control pills prevent you from ovulating. For this reason, strippers on the Pill aren't tipped as well as their colleagues.

Of course, don't expect your lover to *know* you're nearing ovulation. Most men are oblivious about these things, at least at the conscious level. That's fine. Let him think it's him, not some hormone, that's making you so hot.

Makeup for Lower Estrogen

Your face is prettier in the middle of your menstrual cycle. Great, you say, but what can you do the rest of the month, or later in life, when estrogen levels naturally drop? The answer is to fake it with foundation and other cosmetics. When the women in Miriam Law Smith's study wore makeup, the raters rated them as just as attractive, feminine, and healthy as when they were in the fertile days of their menstrual cycles. The researchers concluded that skillfully applied maquillage can compensate for or conceal natural cues relating to hormone levels. Think of cosmetics as women's way of fooling men into thinking we're at our fertile peak when we're not. (There's a reason why cosmetics have been found in archaeological digs of the earliest human civilizations.) Eye makeup makes the eyes look bigger, lipstick makes the lips appear fuller, and foundations and moisturizers make the complexion glow.

Why are you sometimes drawn to macho guys even if they're not your type?

Suppose you're a gal who usually goes for big-hearted, big-brained guys—the physicist, the investment banker, the political organizer, faithful guys you could imagine as fathers of your children. You value men who are assertive without being too aggressive and good-looking but not necessarily macho. Other qualities are more important. But every once in a while, even in the embrace of a committed relationship, something totally unexpected happens. You get weak-kneed over a man who's not at all your type. He's a player, the type who loves and leaves. Or he's your strong-minded professor or boss's boss. His smile is domineering. He calls you baby and touches you on the shoulder. You give him a teasing glance and then wonder, is this you?

A likely answer is yes; it's you on the hormones that precede ovulation. Not only are your estrogen levels higher at this time, which improves your appearance, you also get a burst of testosterone, which keys up your sex drive. (Couples have sex 24 percent more often when the woman is in the five-day window around ovulation than at any other time of her cycle.) At this time, at least on a subconscious level, you might pay more attention to high-testosterone cues: deep voices, broad shoulders, masculine bodies, strong jaws, and dominant behavior. Baby-faced Pete Doherty or

Know Your Pushes and Pulls

If you have a partner, you might be predisposed to cheating on him on a monthly basis. At least this was true of the women who participated in a study by biological anthropologist Elizabeth Pillsworth and psychologist Martie Haselton at UCLA. Female participants had been in a long-term relationship for at least fifteen months. During the fertile part of their cycles (right before and during ovulation), the women who thought their own partners were not very good-looking were more likely to desire other guys, express a desire to meet other men, and consider an affair, even if they were completely satisfied with their relationships. It appears that once women are in a stable relationship and long-term needs are met, they may be more inclined than single women to be attracted to the macho, domineering type. (The very same results were found in a Czech study.) Even more mysteriously, the bland-looking boyfriends of the female participants developed a clever counterstrategy: smothering their partners with love. Women reported that their partners were sweeter, more loving, and more attentive during the fertile phase of their cycles than at any other time. (See page 43 for a possible explanation, and page 302 on mate guarding.)

Leonardo DiCaprio might pale in contrast to a macho Michael Jordan or Tom Brady type.

To test whether ovulating women are more attracted to very masculine men, evolutionary psychologists Steven Gangestad and Christine Garver-Apgar at the University of New Mexico recruited nearly eighty undergraduate men, who were under the impression they were in competition for a date with a hot chick. Each guy was videotaped as he talked about himself and told his competitors why the woman should choose him. The researchers then recruited more than four hundred undergraduate women to watch video clips of the men and judge them on their personal qualities and their desirability as long- and short-term partners.

The results confirmed a well-established pattern found in many studies: ovulation has a powerful effect on women's behavior. The women were consistently attracted to the men with traits of long-term partners—warm, faithful, financially successful, and with paternal potential. However, women who were in their fertile phase at the time of the study gave higher-than-usual ratings to swaggering, competitive, confrontational, dominant-looking and -acting guys, the type of men who look as if they could get their way. The closer to ovulation they were, the more likely these women were to appreciate "tough-guy" qualities in short-term mates.

Concrete evidence that hormones influence women's attraction to high-testosterone men comes from a subsequent study by psychologist James Roney and his colleagues at the University of California at Santa Barbara. Roney measured the testosterone levels of thirty-seven men and the estradiol (estrogen) levels of sixty-five women and asked the women to rate the attractiveness of the men. Sure enough, the higher the woman's estradiol level, which coincides with ovulation, the more attracted she was to the men who had high levels of testosterone. High estrogen loves high testosterone.

Oddly, if you're already in a committed long-term relationship, there may be a better chance that you'll go for ladykiller types than if you're single. At the Face Research Lab at the University of Aberdeen, women were asked to digitally "masculinize" or "feminize" men's faces according to their physical ideal. Among the women who were ovulating, those who had partners made the ideal male face more masculine, whereas single women made their ideal male face more feminine (nurturing).

Intriguingly, a committed woman's attraction to strong, masculine features was particularly strong if her own partner lacked them. Studies have also found that around the time of ovulation women are less attracted to their partners if those guys are asymmetrical, and more attracted instead to the faces and body odors of dominant-looking and -acting symmetrical guys, probably due to testosterone-derived compounds in their sweat. This is true even of women who say they're perfectly happy with their relationships.

This push-and-pull between devoted long-term types and macho short-term ones sounds loony, but from an evolutionary perspective it makes sense. Researchers speculate that the reason women are unconsciously drawn to strong-jawed, deep-voiced, domineering, healthy men when they're most likely to conceive is that a child born of the union inherits the man's "good genes." According to evolutionary psychologists, women associate high-testosterone men with developmental fitness because testosterone is an immune suppressant and only a healthy body can thrive in the presence of the powerful hormone. A father's contribution of genetic fitness to offspring is especially important in the context of a short-term relationship in which he wouldn't necessarily help raise the child. (No one can be certain about family structures in ancestral environments, but it's possible that women raised children with the help of an extended family and/or nurturing partners who weren't necessarily genetically related to her kids.) As we all figure out sooner or later, there are strong, silent, dominant types who wouldn't make marvelous companions. All that testosterone is linked with aggressiveness, irritability, philandering, lack of interest in fatherhood, and a dearth of emotional warmth.

Of course, women generally desire that rare best-of-both-breeds partner who is warm, faithful, intelligent, and cooperative *and* has high-testosterone attributes such as social dominance and confidence (without cockiness) and a good physique. Judging by his looks and reputation, the actor Brad Pitt exemplifies this marvelous hybrid. Brad has a manly jaw and strong cheekbones. He also has big blue eyes, long eyelashes, and plump lips. He looks nurturing and cooperative, yet also very masculine, which fits his public persona as a baby-stroller-pushing dad who plays bad boys in the movies.

According to evolutionary theory, if you can't get a Brad Pitt type, you'll have to compromise. For a short-term relationship, you might have a fling with someone like him, or an even more masculine guy, especially when you're ovulating. For a long-term relationship, you might lower your standards to find the best combination of qualities in a partner (see page 237 for trade-offs). A minority of married women, finding themselves with a mate who falls below expectations, have affairs. While cheating isn't the norm, it's more common than you might think. Around one out of every five married women has confessed to cheating at least once on her husband.

Bottom line: the next time some playboy hits on you and you feel his pull, take a look at the calendar. It could be his musky smell, his swagger, his smoldering gaze, or his husky voice that calls to you—but it's more likely his good timing.

Why don't people go into heat like other animals?

You might look, feel, and act sexier around the time you're most fertile, but your behavior around the time of ovulation is subtle compared to that of other animals. When other female primates enter their fertile phase (estrus, or "heat"), they vocalize their status, redden in the rump, grab at penises, and allow themselves to be mounted by the first male that comes around. No lady does any of the above. But why not?

The answer is that we do something that may seem even

more promiscuous—we have sex all the time, at any time of the night and day and month, all year round. Men don't have to wait for a woman to be in heat to get some action. The human body is designed for more or less constant sexual receptivity. From a guy's perspective, this sounds great. Believe it or not, it's actually even better from a woman's perspective.

Concealed (or nonadvertised) ovulation, according to biologists Richard Alexander and Katherine Noonan, favors women because it promotes long-term relationships. We have sex for both recreation and reproduction. All that lovemaking strengthens the bond between you and your sweetie. And because guys don't know when you're fertile, they're more likely to be faithful, unlike other species in which the males jealously guard females when conception is possible, and sniff around for other fertile mates the rest of the time. Women continue to have sex even when pregnant, which further reinforces their relationship with their partners. All this sex is good for the woman and the baby because it facilitates bonding and means the man is likely to assume the kid is his and help provide for the family.

Anthropologist Sarah Hrdy at the University of California at Davis put a different spin on concealed ovulation. The gist of her theory is that by not advertising when she's fertile, a female can mate with many males of her choosing. If her ovulation were advertised and she showed signs of being in heat, as gorillas do, dominant males might guard her to prevent others from impregnating her. If no one knows the state of a female's uterus, no one knows when to monopolize her. This theory suggests that early human social structure was very polygamous. Even if females had primary mates, concealed ovulation would have made it possible to cheat on that mate with other males who had more desirable qualities (and diversify the gene pool if the new male was an outsider). Considering how male chimpanzees, gorillas, tigers, and other mammals commonly kill babies not genetically related to them, Hrdy concluded that concealed ovulation protects infants. When paternity is unknowable, any male could be a father, and therefore has little incentive to kill infants, and more incentive to invest in them.

So what's the right answer—do women conceal ovulation to keep men around (monogamy) or to have sex with many of them (polygamy)? Biologists Bigitta Sillén-Tullberg and Anders Møller found a nice compromise, as evolutionary biologist Jared Diamond points out in his book *Why Is Sex Fun?* In the course of evolution humans may have swung from primarily polygamous to primarily monogamous (or, at least, less promiscuous). When our prehuman ancestors were wildly promiscuous, it was in the female's best interest to conceal ovulation. Later in the course of evolution, when there was relatively less swinging going on, concealed ovulation remained useful because it promoted male-female bonding (see pages 295–297 for further details).

You can argue that humans are still a bit of both—promiscuous and deceptive, and also faithful and familial. Women's temptation to cheat on their mates around ovulation time and men's ability to detect subtle ovulation cues are signs that the sexes are still battling it out.

Take an Ovulation Test

If you're trying to get pregnant or just curious about when you ovulate, the only reliable way to know is by using an ovulation predictor kit (OPK). Found in most drugstores, these inexpensive pee-on-a-stick tests detect luteinizing hormone (LH) surges. Approximately 24–48 hours before ovulation, your pituitary gland secretes a burst of LH, in concert with another reproductive system hormone, follicle-stimulating hormone (FSH). Generally speaking, the LH surge triggers the release of the egg from the ovary. Test up to a week before the midpoint of your cycle for a two-day advance warning of ovulation. But be forewarned: as birth control, ovulation test kits are not so effective. Sperm can live inside a woman's body for up to five days, which means the sperm that fertilizes your egg may be inside you, hanging out, even before you can detect an LH surge. But for women who are trying to get pregnant, or for those who'd like to understand their cycles better, the OPK is A-OK.

How do the seasons affect your sex life?

Humans have sex all year round, all the time, but that doesn't mean seasonal shifts don't slyly affect us. During autumn in the Northern Hemisphere, men's and women's testosterone levels rise, according to several studies, including one led by endocrinologist Sari van Anders at Indiana University. A Danish study on seasonal variation in testosterone also found evidence for a summer peak in men. Despite disagreement on the exact timing of the peak—perhaps due to cultural and/or geographical variation—the general trend is that levels of circulating testosterone are higher in fall and, perhaps in men, the summer as well. Testosterone levels are associated with sex drive: the higher, the hornier.

Why would people have higher testosterone levels in the fall and summer? No one knows for sure, but van Anders and other endocrinologists suggest that an increase in testosterone levels could be triggered by rapid decreases in daylight. The waning hours of sunshine at the end of the summer and fall may tap into old mating instincts. After all, other animals living in the Northern Hemisphere—including deer, elk, woodchucks, some species of birds, and the Alaskan mountain goat—have their breeding seasons in the autumn. The hypothalamus, the region of your brain that releases sex hormones and responds to pheromones, also happens to be sensitive to light and circadian rhythms. Another reason testosterone levels have been found to be higher in fall and summer could be increased physical activity during those seasons. Exercise, social interaction, time of day, mood, and sex all affect our circulating hormone levels. Urbanites—or anyone else spending most of the time indoors under artificial lights—might not experience such seasonal variability because they're not as pegged to cycles of natural light.

There's no concrete proof that the libido rises as the leaves fall, yet there's circumstantial evidence in the birth record. In the United States more babies are born in August than at any other time of year, followed by July and September, meaning the most conceptions take place in the autumn, nine months earlier. In Europe, births tend to peak in the spring, meaning more conceptions in the summer months. Interestingly, STDs also peak in the summer and autumn, according to studies in the United States and Europe.

Pregnancy aside, a woman's body subtly changes shape as the seasons change, as van Anders observed of the subjects in her study. That's because patterns of fat deposition vary with testosterone levels. In women, high testosterone is related to a high waist-to-hip ratio (a thicker, more tubular figure). Come summer and autumn, you might notice that your body isn't necessarily its bikini best and your pants feel a little snug at the waist. Weight gain is not to blame; it's the redistribution of body fat to the abdominal area. In men, it's the reverse: the higher the levels of fluctuating testosterone, the lower the waist-to-hip ratio, meaning they have slightly less flab around the waist at beach time.

So what does this mean for your love life? Tennyson wrote, "In the spring, a young man's fancy lightly turns to thoughts of love." That might be true. But in the summer and fall there's a fair chance it turns to sex.

What can you tell about people by the ratio of their fingers?

Grasping your date's right hand, turn his palm to face you and look at his fingers. Which is longer: the ring finger (next to the pinky) or the index (pointer) finger? Many men have ring fingers that are longer than their index fingers, known as a *low digit ratio*. A low digit ratio is related to how much prenatal exposure a person has had to the hormone testosterone. Generally speaking, the lower the digit ratio, the more prenatal testosterone.

Let's say your date's ring finger is drastically longer than his index finger. (The average man has a ring finger that is only 4 percent longer, so the difference may be subtle.) This probably means his fetal brain was soaking in a lot of testosterone. According to studies by evolutionary psychologist John Manning and his colleagues at the University of Lancashire, and many others, men with low digit ratios tend to have certain traits: assertiveness, good spatiovisual ability, high numerical but low verbal aptitudes, and musical talent. These strong, silent types are also more likely to have raging libidos, higher-than-average sperm counts, and more masculine-looking facial features such as rugged jawlines, more prominent eye ridges, and thinner lips.

Men whose ring fingers are longer than their index fingers are also more likely to be great athletes. In study after study, they perform better in sports such as soccer, endurance running, skiing, and dancing, and might also be driven to push themselves more and compete aggressively. Prenatal testosterone helps build more efficient hearts and vascular systems and stronger musculature. Unfortunately, there's a downside: these manly men may be too cocky, too confrontational, and have too many sex partners.

While many women (and gay men) have ring and index fingers that are about the same length, a woman may also have a low digit ratio. These gals are more likely to be "butch" lesbians, to be jocks, and to have a higher sex drive and more sex partners, a vicious jealous streak, and attention deficit hyperactivity disorder. Like men with low digit ratios, high-testosterone women are likely to be more competitive and assertive than the norm. A friend of mine who has a low digit ratio is tempestuous, sexually voracious, an "alpha female"—and a mother of four, if you don't include her even-fingered husband, whose life she runs.

People who have ring fingers that are significantly shorter than their index fingers, usually women, may have been exposed to higher levels of prenatal estrogen. A *high digit ratio* is associated with good verbal fluency, small waists, big chests, plumper lower lips, slim noses, and narrow jaws. It's also associated with breast cancer—potentially an effect of lifelong high estrogen levels.

Before you give your date a thumbs-down, remember that

these are only statistics. A fit, fertile, musical, numerical, low-digit-ratio hunk is not necessarily going to fight with you, run off with someone else, and leave you with nine howling babies. The influence of genetic and environmental factors on behavior is so complex that it's impossible to point fingers at any one cause.

Measure Digit Ratios

Some people have ring and index fingers of extremely different lengths, which are easy to eyeball. In others the differences are subtler and require careful measurement. To accurately measure digit ratio, use a ruler or, better yet, vernier calipers (available at hardware stores). Measure the index finger of a person's right hand from the bottom crease near the palm up to the fingertip, placing the measuring device directly on the flesh of the finger. (Photocopying or scanning your hand may flatten your finger pads and throw off your measurements. Particularly for men, the right hand is more affected by prenatal hormones than the left.) Repeat the same for the ring finger. Divide the length of the index finger by the ring finger. The typical ratio is .96 for men (ring fingers are slightly longer) and 1.0 for women (index and ring fingers are the same length).

Why do men lose their judgment and decision-making skills when looking at pretty women?

Imagine that two macho guys are told to split $100. The first guy's role is to suggest how to divvy up the money, and the second guy gets to accept or reject the offer. If the second guy accepts, the money is distributed according to the agreement and both men go home with cash in their pockets. If the second guy rejects, both go home empty-handed.

When this negotiation, known as an ultimatum game, is played in a laboratory setting, men with high testosterone levels are known to furiously reject unfair offers (anything below a 50/50 split). They almost always prefer to punish the other player, even if it means they must go home with nothing in their pockets. But not always. Belgian psychologist Siegfried Dewitte and researcher Bram Van den Bergh reversed that trend when they showed male players a picture of a sexy babe before the game. When high-testosterone guys in their twenties and thirties saw the picture of the woman, they were much more likely to accept unfair offers (less money) than their peers who hadn't seen her. Distracted, they were willing to just take the money and run.

In a similar study at McMaster University in Canada, experimenters offered players a choice between a small sum of money the next day or a larger sum later on. The long-term payment was a substantially bigger payout, and almost everyone chose it over the smaller short-term payment. Only guys who had been gaping at photos of beautiful women settled for less money sooner.

Why do men shortchange themselves when they look at babes? The answer is that arousal overrides the part of the male brain that makes rational decisions. According to the researchers, men evolved to respond swiftly to sexual cues. Men's testosterone levels are known to shoot up as much as 30 percent after an encounter with an attractive woman. Looking at a gorgeous female face also floods the reward centers of their brains with the feel-good neurotransmitter dopamine.

Triggering this knee-jerk reaction is the amygdala, the brain region associated with emotions, urges, and spur-of-the-moment decisions. When a guy's amygdala is hyperstimulated, he reacts strongly and impulsively. The larger this almond-shaped area, the stronger the sex drive. The stronger the sex drive, the more aggressive the response to sexy stimuli such as pictures of gorgeous women. (No shock here: men's amygdalae are proportionally larger than women's.) For high-testosterone men in particular, whose amygdalae are larger and sex drives higher, short-term gratification (responding to a sexual cue) trumps long-term gain. Imagine a teenage caveman impulsively dropping his sack of meat at the sight of a receptive woman, and you get the picture.

The volatile combination of high levels of testosterone, dopamine in the reward regions of the brain, and a large, hyperstimulated amygdala also explains why cocky young dudes so often make dumb decisions in the presence of attractive women—they drive too fast, spend too much money, and have unprotected sex. The phenomenon hasn't been lost on strip club and casino owners, who get rich by employing sexy, scantily clad beauties to serve drinks to loaded patrons. As my friend Ruben, a high-stakes gambler and Wall Street trader, put it, "You gotta ask yourself, what are you doing? If you're trying to focus on a bet or a trade, you can't be looking at girls. Trust me, you'll lose every time."

Clearly, for high-testosterone men, the subconscious decision to focus on the present seems the better strategy. Why presume that riches tomorrow are better than booty today? Carpe diem! These are guys who'd rather be shortchanged than undersexed.

Beware Overaroused Men

Has a guy ever told you he loves you in the heat of the moment only to take it back later—as in after you have sex with him—leaving you emotionally pulverized? Behavioral economists Dan Ariely at MIT and George Lowenstein at Carnegie Mellon explored how sexual arousal disastrously affects men's judgment and decision making. They asked male undergraduates to answer questions about their ethical boundaries and behavior in one of two physical states: either aroused (by masturbating to pornography) or unaroused.

Disturbingly, the sexually aroused men—answering

questions with one hand on a Saran-wrapped laptop keyboard and the other on their erect penises, with an image of a near-naked woman on the screen—said they'd be significantly more likely to find it fun to tie up a sex partner, imagine being attracted to a twelve-year-old girl, tell a woman they love her to get her to have sex, keep trying to have sex after a date says no, get her drunk to have sex with her, slip her a drug to increase the chance she'll have sex, and imagine getting sexually excited by contact with animals. (Interestingly, sexual arousal didn't make straight men more compelled to have sex with another man.)

The researchers call this phenomenon the "hot-cold empathy gap," meaning that when men are not aroused, they act more rationally and empathetically—and underestimate the impact of being aroused. According to Ariely and Lowenstein, men's self-control when sexually aroused comes not from willpower, which probably wouldn't work, but from "avoiding situations in which one will become aroused and lose control." To avoid the possibility of date rape, the researchers suggest that women be aware that, although it's no justification, sexual arousal perversely affects men's decision-making skills and makes them more likely to engage in morally questionable behavior.

Why does doing something dangerous or exciting increase attraction?

Call it love at first fright, although "excitation transfer" is the scientific term. It's what it sounds like: a transfer of excitement from one experience to another, most often from danger to sexual arousal. When excitation transfer happens, it's like having the Midas touch—but instead of turning to gold, everything just turns sexier.

Psychologists started to pay close attention to excitation transfer in the 1970s, after the famous "bridge" experiments at the University of British Columbia. Two psychologists, Donald Dutton and Arthur Aron, instructed a pretty woman to approach men who were crossing the Capilano Suspension Bridge. The bridge is a well-known thrill spot: a creaking, swaying span of wooden planks, attached to thin wire cables, neck-breakingly high above rocks and rapids. The woman, who wasn't aware of the true purpose of the experiment, asked male bridge crossers if they'd participate in a project for her psychology class on creativity and scenic attractions, and to write a brief story based on a picture. She gave her name and number to each participant and invited him to call if he wanted to talk again. The same procedure was also done on other men in tamer settings—over a stable bridge, and in a park where men were relaxing after having crossed the dangerous bridge.

When impartial judges read the men's stories, they rated those written by the bridge crossers as more sexually explicit than

Create Your Own High

Reading the wedding announcements in the newspaper, I often come across the strangest stories of how couples meet: a car accident, a hurricane, on a battlefield in the line of fire. During a blackout emergency in Manhattan, at least two people I know had one-night stands with near-strangers. Similarly, a female friend had a fling with her instructor at a surf camp in Hawaii; he was a leathery tough-guy type she'd otherwise never give a second glance. A guy friend who met his wife at a club admitted that he's definitely more attracted to women in exciting settings such as sporting events and parties, when he's transported by a rush of people and noises and smells and colors. Now it all makes sense. The residual excitement from an invigorating or even dangerous experience can boost a person's sexual attraction to a stranger, or perhaps jump-start an existing relationship that has lost its energy. Meet your match while skiing, rafting, surfing, parachuting, hang-gliding, mountaineering, deep-sea diving, surfing, storm chasing, or doing anything else that's a stretch for you and stimulates your brain and heart. It will get your cheeks blushing and your heart racing . . . and you won't be the only one.

those written by men in safer settings. But even stronger proof of excitation transfer came by phone. More than 50 percent of the Capilano veterans called the woman compared to fewer than 25 percent of the men from the other settings. The bridge crossers, in their high, simply found the woman more attractive.

Many years later, psychologists at the University of Texas at Austin did a similar study. They asked three hundred thrill seekers who were either standing in line to ride the roller coaster or on the way out the exit to rate a photo of a good-looking person of the opposite sex. The riders also were asked to rate their seatmate's attractiveness and to identify their relationship to that person.

Supporting the theory of excitation transfer, the men and women who were approached after the ride, when they were shivering with excitement, rated the strangers in the photos as substantially more attractive and desirable as dates, whereas the people who rated the photos before the ride gave the strangers lower attractiveness ratings. Again, the high heightened sexual desire.

It's worth mentioning that roller-coaster riders weren't inclined to give their partners higher ratings after the ride than before the ride. It could be that excitation transfer works only on unfamiliar people, in the absence of any information other than their looks. Or it could be an outcome incidental to the experiment—people gave their sweethearts high attractiveness ratings before the ride as well as afterward—especially considering that those sweethearts were nearby and might look over their shoulders. Another possibility is that a rousing scare might make you and your partner more affectionate and intimate instead of more attracted to each other's looks. (Think of all the canoodling that couples do after or even during emergencies such as blackouts and blizzards, or make-up sex after a big fight.)

Why is excitation transfer so powerful? The answer is that the neural pathways that are engaged in an arousing stimulus overlap with those of sexual arousal. In these moments, your mind and body are charged from the effects of dopamine, a hormone that stimulates your brain's reward areas and boosts your mood. Add to this adrenaline, which gets your heart pumping as it increases the supply of oxygen and glucose to your brain and muscles, and testosterone, which stirs the libido. The rush from an experience may last for ten minutes or so, occasionally for a few hours, and may enhance, intensify, and eroticize everything you encounter.

As with any arousing experience, you owe the sensation of excitation to the amygdala, the region of the brain that produces and responds to emotions and is related to the strength of the sex drive. The amygdala is superstimulated when you do anything stirring, daring, or novel. (Yes, there's a fine line between excitation transfer and overarousal.) Thank this almond-shaped cluster for some of the sexiest, most memorable experiences of your life, and blame it for the bad memories that you can't forget. It plays a crucial role in processing an experience and deciding whether and where and how to store it as a memory. Bearing in mind that excitement stimulates the amygdala, you can understand why the gal on the Capilano Bridge left such a lasting impression on all the men who encountered her.

It all goes to show that to find the right person, you need only be at the right place at the right time, even if it's on a shaky bridge. Remember, love itself is a suspension of disbelief.

Why do you like and trust a guy more after you've been intimate (even just cuddling)?

If all you want is pure, physical, no-strings-attached sex, you're up against a lot more than the establishment. The enemy is within, and it's your hormones. Stroking, cuddling, hugging, kissing, and having sex—any intimate activity willingly given or received—can trigger hormones that dissolve your boundaries.

Betraying you is oxytocin, aka the "hug drug" or "cuddling chemical," a hormone known for its role in mediating close emotional relationships. Oxytocin works by suppressing the activity of the amygdala, the part of the brain that processes fear and triggers actions. It helps you get over your aversion to strangers. At the same time, it stimulates receptors in the hypothalamus, the part of your brain responsible for the release of sex hormones and unconscious activity such as heart rate and blood pressure. Oxytocin also scales back the production of stress hormones and lowers blood pressure. (Don't confuse it with the powerful painkiller Oxycontin.)

You know that warm surge that passes through your body when you snuggle with someone? It's probably oxytocin, breaking down your guard and helping you feel closer, calmer, and more connected. When researchers sprayed the "love hormone" up the noses of people with social phobia, subjects became less anxious and more sociable. In one study, men who inhaled six puffs of oxytocin showed significantly higher levels of trust in an experiment that involved transferring money. Trustworthiness follows a virtuous cycle—the more oxytocin the more trust, the more trust the more oxytocin, and so on.

Now imagine what oxytocin can do to the horny brain. More touching leads to more trusting, and more trusting leads to more touching. The intensifying cycle of touching and trusting could well lead somewhere. In a sexual context, oxytocin works in tandem with other neurotransmitters, such as testosterone, that fuel your desire to make love. Oxytocin may also influence how the feel-good neurotransmitters dopamine and norepinephrine hit the reward parts of the brain. This killer cocktail of trusting oxytocin, horny testosterone, and passionate dopamine explains why, to your horror and against every shred of your will, you find yourself getting too emotionally attached to your married lover or the immature dude you picked up last weekend. After a weekend sexfest with her ex-boyfriend, a friend of mine gushed that she was falling in love with him again. She looked positively coital. I assured her that she wasn't really in love—she'd dumped the guy years before, and I knew she didn't want to go back to their troubled relationship. Sure enough, the effects of the oxytocin wore off within a week after he went back home to another city. She went through a minor withdrawal, then carried on.

Women experience stronger effects of oxytocin than men

because women have more estrogen, and estrogen makes oxytocin receptors more sensitive. When your estrogen is down, so are the effects of oxytocin, which may be another reason why you respond more strongly to touching during the high-estrogen ovulatory phase of your cycle and less strongly during menstruation and after menopause. (For men, the neurotransmitter vasopressin has a greater effect on bonding; see pages 304–306.)

Both men and women enjoy a thirty-second oxytocin surge during orgasm, but don't for a moment think oxytocin's job is finished when you're spent. Somewhat mystically, the hormone increases your chances of pregnancy by facilitating the postorgasmic contractions that pull sperm toward the egg. Later on, it induces labor contractions, helps you bond with your baby, and gets your breast milk flowing. Oxytocin is a sex coach, fertility facilitator, doula, and wet nurse all in one.

All this said, a little cuddling isn't going to make you too attached to your masseur. The hormone may take a while to build up, at least in some people. In a study on rats, it took two weeks of regular daily massage for the rodents to show general increases of oxytocin in their systems. The truth is, everyone has different capacities for producing and processing it. If you're used to a lot of touching and hugging, your brain may produce oxytocin easily. If you were born and raised in a family without much touching or hugging, you might not produce or process as much oxytocin (a reason why people who didn't get touched much as babies are often less trusting and have intimacy issues). If you're stressed or have PMS, your brain blocks receptors for oxytocin; you might not even want to be hugged. Nevertheless, getting affection from someone generally helps you give it, and giving it helps you get more of it.

But first make sure you want it.

Get Your Oxytocin Fix

How do thee get an oxytocin boost? Let me count the ways. Naturally, any bodily contact with someone you like might do the trick. Nipple stimulation and orgasm are considered especially effective. Acupuncture and massage are also said to release oxytocin in your system. The same goes for the club drug Ecstasy, which makes even the surliest teen want to cuddle (I'm just reporting here, not endorsing). Nonphysical activities may work, too. Simply donating to charity has been shown to significantly raise oxytocin levels.

Sensation seekers and the very shy are tempted to buy sketchy "trust-in-a-bottle" sprays online, or ask their doctors for a prescription (trade names Syntocinon and Pitocin), which is often used to induce lactation in new mothers. There's no consensus on how well synthetic oxytocin works for social purposes, how safe it is, how much you need, or for how long. Once in your bloodstream, the chemical is likely to deactivate within an hour or so. A more consistent and reliable oxytocin source is a loving partner.

Why do men mellow out when they're in a relationship?

"She's good for him," I observed when a friend of mine, Rob, a self-described brute, became a teddy bear when he fell in love. Yes, a man who had been a belligerent and combative bachelor actually became something of a nice guy after he got married to a cheerful, lovely woman. When I pointed out to him how much his personality had changed for the better thanks to such a kind and supportive partner, he grunted and gave me a half smile. "Just regular sex," he said.

That's probably part of it, but there's more. A likely reason Rob is sweeter is that he has less testosterone circulating in his body than when he was single. In a study of 122 male heterosexual Harvard Business School students, those in committed monogamous relationships had testosterone levels 21 percent lower than those of their single classmates. It's not as if committed guys have no *cojones*—behavior can't be totally reduced to hormone levels. But men with lower testosterone levels are on the whole less aggressive, less irritable, and less likely to take life-threatening risks or drop dead of a heart attack. (No wonder married men live longer.)

The Harvard study's authors, Terry Burnham, Peter Gray, and their colleagues, point out that the lower testosterone levels are seen only in men who are genuinely committed, not in those on the prowl. The anthropologists conducted similar studies in places such as Kenya and Beijing where married men commonly have

Test His Testosterone

Fluctuations in testosterone levels don't have an extreme effect on overall behavior as long as they're in the normal range (300–800 nanograms per deciliter of blood), but levels may surge dramatically in as little as ten minutes if a person is stimulated or angered. If men's levels are always too low, they lose their libido and feel sluggish. If they're always too high, they may suffer from prostate problems, injuries, and diseases associated with riskier behaviors. High-testosterone guys may also suffer from an inability to be happy in a long-term relationship, which means their partners suffer, too. If you and your partner are worried that his testosterone levels are abnormal, see a doctor for a hormone test or order a do-it-yourself home test kit. (Note: California and New York prohibit home testing for hormone levels.)

lovers on the side or spend less time with their partners and more time with their buddies in bars and at sporting events. Marriage didn't affect the testosterone levels of these men; they were the same as those of bachelors. Husbands who are cheating or want to cheat also have higher testosterone levels. It's as if these dudes are still competing in the mating market.

What comes first, you might ask—a committed relationship

or lower testosterone? The answer is that it likely works both ways. There are men who had been exposed to high levels of prenatal testosterone in the womb. Additionally, there is a level of circulating hormone in the body, which rises or falls depending upon context. Both of these variables may affect behavior. While men with lower baseline testosterone levels are more likely to be in loving, committed relationships in the first place, relationship status may also affect circulating hormone levels. Behavior drives hormone levels just as hormones drive behavior.

For men and women alike, testosterone levels jack up when we feel the need to assert dominance or status, and dip when we don't. Holding a gun can trigger a testosterone rise, and so can winning a chess match. In a study of twenty-one hundred Air Force veterans, sociologist Allan Mazur discovered that men's testosterone levels rise after a breakup or divorce, drop when the men remarry, and remain stably low for the duration of the relationship. Single men may have higher testosterone levels because they must compete with other men for mates and because they lack the calming support of long-term girlfriends or wives. Stability, security, and regular, unthreatened sexual access—these factors alone might make men's testosterone drop. But beware the man whose overall levels remain high. In Mazur's study, husbands who maintained levels of testosterone one standard deviation above the mean were 43 percent more likely to get divorced than those with normal levels, 31 percent more likely to leave home due to marital problems, 38 percent more likely to cheat on their wives, and 13 percent more likely to report that they hit their wives or threw things at them.

If it's not enough that a man's testosterone level drops when he commits to his spouse, the effect becomes even more dramatic in a different long-term relationship: fatherhood. The postpartum plummet occurs in dads around the world, from America to Africa. (As my brother put it, fatherhood made him a "wuss.") According to a study by psychologist Anne Storey, testosterone levels were 33 percent lower among new and expectant fathers, and holding a baby or even a baby doll makes testosterone levels tumble. Spending time with babies, not babes, does the trick.

There's a probable evolutionary reason why committed men, especially fathers, have lower testosterone levels (and different levels of other hormones as well; studies are under way on vasopressin, which facilitates bonding, as well as estrogen and prolactin). A lower-testosterone man is more sympathetic and more likely to pay attention to his children. There's a better chance that he's a faithful companion, and less likely to drop dead in a fistfight or from heart disease, prostate cancer, or other diseases linked to high testosterone. Which is a good thing, because it means he'll be around to change the diapers and help pay for college.

When are you most committed to your partner?

Throughout her twenties, a good friend of mine, engaged to her long-term boyfriend, would find herself of two minds about her relationship. We would have lunch, and the conversation would turn to their impending wedding and the dream house they wanted to buy. She was genuinely excited. But there were other times when the talk would take a sideways turn, and somehow we'd end up reminiscing about a sexy fling she'd had ten years before, or her eye would fall on the hot guy nearby, pounding away at his laptop. She'd turn to me, biting her lip, and say something like, "I love him so much, but sometimes I feel like my heart's not in it. Do you know what I mean?" I'd nod. She'd sigh and shrug. Of course I knew what she meant.

Such are the ups and downs of relationships, and they may be in part driven by hormone cycles. Women's hormonal profiles are similar in the luteal phase of the cycle (between ovulation and your next period) and during pregnancy. At these times your body has relatively low levels of the "sexy" hormones estrogen and testosterone. At the same time, you have elevated levels of the hormone progesterone, which rises sharply after ovulation and dips right before your period. Not much is known about the effects of progesterone in humans, but in animals it triggers bonding behaviors that precede mating and maternal bonding. It may also reduce anxiety and make you more attentive—which doesn't hurt a relationship.

To test the luteal-phase commitment theory, evolutionary psychologist Ben Jones and his colleagues at the Face Research Lab at the University of Aberdeen recruited nearly a hundred women in committed relationships and asked them to complete diaries and take tests throughout their menstrual cycles. They also showed the women twelve pairs of male and female faces that had been digitally manipulated, one masculinized (stronger cheekbones, squarer jaw, etc.) and the other feminized (rounder, softer features), and asked them to choose which of the two versions they found most attractive.

It turned out that the women in the luteal phase of their cycle preferred soft-faced men more than women who were in their fertile phase, who were more attracted to the hard-faced toughies. They also gave nurturing-looking women higher ratings. Strikingly, women in their luteal phase also reported increased commitment to their partners, even though all the women in the study said they were happy in their relationships.

Jones and his colleagues speculate that when progesterone levels are high, women strengthen their relationships with their partners, family, and friends. The hormone may facilitate an innate bonding reflex. Our foremothers, when pregnant and lactating, would have benefited from a large social circle, including mothers, sisters, aunts, and brothers. A subsequent study at the Face Research Lab found that women in their luteal phase are also more attracted to faces that look like their own (kin).

It follows that bonding might not happen as naturally in your fertile phase. When your estrogen and testosterone levels are high, you're attracted to high-testosterone cues, not bonding and nurturing ones (see page 122). You're more likely to compete with other women, including friends, for mates. Seen another way, we might normally bond well with our partners, friends, and family—except when our hormonal profiles subtly shift between menstruation and ovulation.

It's a fascinating idea—that, on an evolutionary level, our

social behaviors are exquisitely timed to hormonal fluctuations that subtly compel us to commit, or compete. These pushes and pulls, the strengthening and weakening of social bonds, have forged our species.

Watch Your Midmonth Mouth

The next time you hear yourself inadvertently pointing out a friend's weight problem or spreading spiteful gossip about another gal's sex life, check the calendar. In a study by Maryanne Fisher, a psychologist at Saint Mary's University in Canada, women who were in the fertile phase of their cycles subconsciously rated other women's attractiveness lower than did women who were in the high-progesterone, low-estrogen phase of their cycles (men's attractiveness ratings, meanwhile, did not go down). Fisher suggests the reason for downrating other women is intrasexual competition and is due to hormone levels: around ovulation, when estrogen and testosterone levels spike, women may get more aggressive and rivalrous. Female friends become competitors for male attention. A woman in the fertile window of her cycle may be more likely to derogate other women when talking to a man, helpfully pointing out that a rival sleeps around or has had a boob job. Watch those claws!

CHAPTER 6
Signs and Signals

Fie, fie upon her!
There's language in her eye, her cheek, her lip,
Nay, her foot speaks; her wanton spirits look out
At every joint and motive of her body.

—William Shakespeare, from *Troilus and Cressida*

What body language do women use to express interest?

Mae West once described herself as fluent in two languages: English and body. Mae knew, and Mae's men understood, that body language says as much as, if not more than, the spoken word. The women who get the most male attention are the ones who know exactly how to express themselves. Some body language is conscious and deliberate, but to many it comes naturally and subconsciously.

No successful encounter happens without eyes meeting first, according to Monica Moore, a psychologist at Webster University. Moore identified three forms of gaze: type I, the room-encompassing glance, is a five-to-ten-second scan of the entire room (during which you may be raising your chin, holding in your stomach, and arching your back to stick out your chest); type II, the short darting glance, is targeted at a specific man and repeated several times; and type III, the gaze fixate, is a direct gaze that lasts longer than three seconds. Direct eye contact may be accompanied by a surprised-seeming "do-you-mean-*moi*?" lift of both eyebrows. Smiling during type III eye contact often seals the deal (see page 159).

Once you're eye to eye with a guy who interests you, your head and neck region become his focal point. Frequent signals at this stage are the head toss and the hair flip, which are usually executed simultaneously. In less than five seconds, you tip your head upward and give your hair a little toss. Some women toss in one serious shake, and others pet and stroke their hair as if it were a cat. If you're not caressing yourself, you may subconsciously rub an object such as your keys, rings, or wineglass. Your fingers might casually find their way to your neck, which you massage as you incline it at a forty-five-degree angle, in a gesture of vulnerability and coy arousal.

The deliberate lip lick also got a high ranking in Moore's catalog of nonverbal solicitation signs. When not speaking, you might part your lips slightly and wet your upper or lower lip. If you're not the subtle type, you might lavishly run your tongue over your lips several times, or apply lipstick in the same bold manner, all while maintaining profound eye contact with your target male. Plumping up your lower lip in a pout is optional. The implicit suggestion is that bright full lips are characteristic of youth and high estrogen levels, not to mention the erotic parallels between glistening, inflamed lips and other body parts.

Subconsciously, you're likely to laugh a lot when you're interested in a guy, and do a lot of nodding. If you like him, your body movements may be more animated; interested women tend to subconsciously flex their arms at the wrist and elbow quite a bit. Then there's "play," and it involves pinching, tickling, or sticking your tongue out, or sitting in men's laps. If you're very bold, you may attempt the skirt hike—a quick jerk of the wrist that raises the hem of your dress or skirt so that more leg is exposed. More common is the solitary dance, gyrating your body in time to the music, while sitting or standing. If the guy's in tune with you, he might pick up on your hint and request a dance, during which you may grasp his hand so that his palm is flush with yours.

If you're in a place where the music or ambient noise is loud (or even if it isn't), you may drop your voice to a purring whisper. This inspires him to move closer so you can deliver the message directly into his ear. Your body may accidentally brush against his. Your hand may graze his knee or thigh, your foot might rest on top of his foot, or your breast might happen to press against his chest. Even if you're not so forward, you're likely to lean toward him if you like him. It's as if we all are subject to the forces of magnetism, subconsciously drawing near those we like and away from those we don't. So imagine yourself as a magnet, with a pull so strong he can't resist.

What's the strongest signal you can use to get someone's attention?

The most important signal a person can send to communicate sexual interest (without being crass) is a smile and a direct gaze. It doesn't, nay shouldn't, be an open, gaping stare combined with a broad, sloppy grin. Even a Mona Lisa look—a slight smile combined with a slit-eyed gaze—could work, just as long as you unambiguously direct it at your target. The smile without the eye contact, or the eyes without the smile, will not suffice. Averting

Send Signals

Attraction is a funny thing. We make a big fuss about physical beauty, but psychologist Monica Moore's research suggests actions may be even more important than looks when it comes to getting guys to approach you. Women who displayed more than thirty-five signals an hour were approached by an average of four men, whereas those with less flirtatious body language often weren't approached at all. The women who got men's attention weren't prettier than those who weren't approached. In fact, unattractive women with expressive body language were approached more often by men than attractive women who did not signal. Only when women's body language indicated interest did men respond enthusiastically, because they liked the unspoken thing that was said.

your eyes, or gazing with a neutral expression, sends mixed messages. The look and smile together enhance your recipient's "feel-good feeling" in the reward areas of the brain. When you smile, your recipient's amygdala, which picks up on emotional signals, instantly detects your friendly cue. When your smile is combined with a direct gaze, you're *significantly* more attractive to the opposite sex than when gazing without smiling, according to a study by

Ben Jones and his colleagues at the Face Research Lab at the University of Aberdeen.

Subconsciously, your smile might clue others into your hormonal status. According to studies by psychologist James Dabbs, a person with a wide, friendly smile that reaches the corners of the eyes is more likely to have lower levels of testosterone—arguably a turn-on for men when the person smiling is female. High-testosterone men and women, in contrast, don't smile as often, and when they do, their smiles are tight and brusque (see page 161 for more on the difference between smiles). However, judges observing the smiles of high-testosterone men in Dabbs' study gave them higher "potency" ratings than those of low-testosterone men—arguably a turn-on for women.

By smiling at a person, you're setting yourself apart, according to Jones and his colleagues. Your expression is a social signal that makes your face light up and tells a person that you've singled him out, if only for one shining moment. From an evolutionary perspective, it makes sense for a man to find you more attractive if you're sending him a signal that you're interested in him. If he approaches you, he might have a better outcome than with another woman who doesn't seem interested.

The lesson here, once again, is that attraction is influenced not only by looks but also by how much you engage the other person. You can be absolutely stunning, but your beauty is enhanced in the eyes of others only when you look at them and send the right signals. The more your eyes direct your smile at your target, the more powerful the impression, and the more attractive you appear.

What exactly makes a smile attractive?

A smile only really works when it's real. Try as you might, the wide, electric grin you flash for the camera or a colleague doesn't do as much for you as the spontaneous one that lights up your face when you're happy or amused. Joy radiates only from a genuine smile.

Paul Ekman, a psychologist and expert on facial emotions, once asked men and women to judge personality characteristics of smiling people in videos. Some of the grins were real and spontaneous, while others were "social smiles," friendly but deliberate. Without knowing why, the judges had a much more positive impression of people when their smiles were genuine, and perceived them as more expressive, natural, relaxed, and pleasant than when their smiles were faked. When Ekman showed raters the real and social smiles side by side and asked them to guess which was which, they could detect the real smile nearly 75 percent of the time. But when only one of the two smiles was shown and judges could not compare, they could only detect the real smile 56 percent of the time. If you think about it, it's true—you can often sense when people are genuinely happy, which makes you more attracted to them. But the more common social smile probably confuses you, making you second-guess yourself.

According to Ekman, the difference between social smiles and expressions of genuine delight involves the subconscious manipulation of facial muscles. In a beauty contestant's or politico's fake megawatt beam, only the muscle that lifts the corner of the mouth (zygomatic major) moves. In a real smile, pleasure touches the eyes by way of the orbicularis oculi, a muscle surrounding the eye that makes the cheeks move up and the skin around the eyes crinkle.

It's difficult to feign an orbicularis oculi smile, because most of us only have control of the inner part of this muscle near the eye, not the outer part. (People who have extended crinkles—crow's feet—radiating from their eyes may have done a lot of genuine smiling in their lives.) Note that the involuntary muscle also makes the eyebrows fold down a little bit in the corners. Genuine smiles are smoother and more symmetrical than fake smiles. They spontaneously erupt on the face and linger only as long as the feeling that inspired them.

Why do guys think you're into them when you're just being friendly?

Not long ago I was chatting with the man next to me on a park bench. Golden sunlight was pouring down. He was a tall guy with a big pink face and an acute interest in word puzzles. We mused about the origin of the phrase "passing strange." Then, abruptly,

Smile for a Boost

A fake smile has the potential to become genuine. There's a feedback loop between facial muscles and emotions, so whatever emotion shows up on your face influences your emotions and vice versa. (This is the same reason gazing into a person's eyes makes you feel more attracted; see page 4.) The theory, known as causal facial feedback, was proven in the 1980s in a unique experiment led by German psychologist Fritz Strack. The researchers told study participants that they were testing ways paraplegics might hold pens in their mouths, and instructed them to grip a pen between their lips in one of various positions, including a forced frown and a forced smile. At the end of the experiment, participants were told to rate the humor of a cartoon. It turned out that the men and women who thought the cartoon was funniest were those who had been forced to smile with the pen in their mouths. The researchers concluded that facial emotions not only modify an emotional experience but also initiate one. Try it yourself, by fixing a grin on your face. It's just as the Buddhist monk Thich Nhat Hanh said: "Sometimes your joy is the source of your smile, but sometimes your smile can be the source of your joy."

the conversation stalled and he gazed at me with an expression of rapt interest. "You're really sexy," he said seriously. I thought, *What did I do?*

Guys have easy triggers. Martie Haselton, an evolutionary psychologist at UCLA, asked more than two hundred men and women to report how often a person of the opposite sex either mistakenly presumed sexual interest or underestimated it. Sure enough, women more often encountered men who *overperceived* their sexual interest and men more often found that women *underperceived* their sexual interest. The higher the man's perceived mate value—that is, his desirability based on looks, past dating history, and reputation—the more likely he was to think women were into him when they weren't. Why is it that men get it so wrong so often?

Haselton and her colleagues, including the evolutionary psychologist David Buss, explain men's self-delusion—their sexual overperception bias—in error-management terms. Basically, errors come in two flavors: false positives, when you think something is true but isn't, and false negatives, when you think something isn't true that is. The researchers conclude that men are like fire alarms—biased toward false positives. For firefighters, a false positive means they rushed to a site where there wasn't actually a fire. For a lovelorn man, a false positive means he hit on a woman who didn't actually have the hots for him. As exasperating and wasteful as false positives are, they're less of a loss than false negatives . . . which, for firemen and romantics alike, mean undiscovered flames.

Men's sexual overperception bias has to do with the selective pressures on both genders. Women are biased toward false negatives, so we're more likely to think a man isn't sexually interested than to mistake his intentions. Falling for the wrong man and getting pregnant by him is a huge risk that could ruin future romantic prospects. Better to underperceive an opportunity than to go for every one that comes along.

Women's choosiness makes men more competitive, and to compete effectively against other men, guys can't waste opportunities. Evolutionarily speaking, men who blow their chances with women are outreproduced by men who don't, so they're more inclined to think there's something there than not. From a guy's (unconscious) perspective, it's better to pursue and lose than never to have pursued at all. As a result, men often get rejected and women often get hounded.

Remarkably, neuroscientists have even found evidence for men's overperception bias when looking at brain scans. In a functional MRI (fMRI) study led by Stephen Hamann at Emory University, men and women were shown sexually explicit pictures of couples as well as pictures of men and women having nonsexual interactions. The neuroscientists observed that the amygdala and hypothalamus were more strongly activated in men than in women when watching the couples having sex. This wasn't a shock: this neural circuit is related to physiological arousal and the release of sex hormones, and men are known to respond more strongly to sexual imagery (see page 194). What the neuroscientists didn't expect is that men's amygdalae and hypothalami also became active, albeit less strongly, when watching the neutral interactions. This suggests that for men, even a nonsexual encounter can have sexual undertones. Tell your macho male officemates to "table a discussion," "think outside the box," "touch base," or "cover your ass," and there's a good chance they'll find an erotic innuendo.

Of course, savvy women aren't oblivious to men's overperception bias, and some may use it to their advantage. As the researchers point out, a gal could benefit from leading men to believe she's sexually interested in them when she's not. If a guy thinks there's a chance you'll have sex with him, he's more likely to do favors for you, boost your ego, and inadvertently make you seem more desirable to other men (see page 266 on wingmen). These men have not only chronic overperception but unfortunate overreach.

What body language do guys use to get your attention?

A few years ago, researchers at Bucknell University and the Ludwig Boltzmann Institute for Urban Ethology teamed up to observe the mating rituals of human males. Observation points were set up in appropriate habitats, which for our species are bars and clubs. At these watering holes males and females mingled in close proximity. They stood upright, were well-groomed, and occasionally displayed nonverbal mating signals. By the end of the study, the researchers caught forty men in action as they tried to attract the attention of women.

Eye contact, the researchers observed, is the most important cue, and men on the prowl scan a room trying to catch women's eyes. They're looking for a gal who focuses on them for a moment, swiftly looks away, and then looks back again (see page 156). Males rarely approach women without significant eye play. If eye contact hasn't been established before an approach, women report confusion and discomfort, usually a precursor to rejection. Women had to make an average of thirteen short and direct glances before men gulped down their inhibitions and made a move. (The researchers didn't observe women approaching men first, although eye contact is technically the first move.)

Between eye contact and the actual approach, men make many other subconscious gestures to rivet a woman's attention. A bold man is likely to make "space maximization movements": either spreading his legs while seated or stretching his arms across the backs of adjacent chairs or other guys. He might hook his thumbs in his pockets or belt loops, so that his hands frame his genital area. Not only do these stances make the guy visually conspicuous, they also establish him as dominant among his male friends. Open body positions convey social power, potency, and persuasive personalities, whereas closed postures such as folded arms and rounded shoulders convey diffidence and a lower position in the social pecking order. Men who eventually approached women made an average of nineteen space maximization gestures beforehand, all while sneaking glances at the target female to make sure she was looking and still interested.

Not every man has the courage to approach a woman or gets the "green light," but many in the study expressed interest with automanipulation signals: stroking the chin, rubbing the face, and scratching the cheek. Evidently, these are the subtle ways men draw attention to their faces, particularly to the beard growth areas that suggest masculinity and develop under the influence of testosterone. These gestures may also be signs of nervousness and anxiety and an effort to displace discomfort.

Women are the ones doing the selecting. Even when talking to your girlfriends at high speed and top volume, balancing both wineglass and cell phone, you can check out a guy on the other side of the room, right? This includes taking in his gesticulation patterns. Men who make frequent hand gestures, especially with their palms facing up, are perceived as having communicative and agreeable personalities that give them greater success with women. Touching is also important. The man who jocularly slaps his friends' backs is perceived as having greater social cachet than the slappee, and women generally find that attractive. Men who reciprocate the back slap are perceived as having higher social status than the guys who don't touch or get touched by other guys. Throw together a group of men and a social hierarchy emerges instantly.

Given all the staring, thigh spreading, backslapping, and body scratching, it has occurred to anthropologists that human males closely resemble apes in their mating body language. Change the habitat from jungle to tiki bar, throw in a few beers, and you'll see how convincingly men can make monkeys of themselves—and how women love it.

How persuasive is a touch?

Want to improve your chance of getting your way? Touch the person who can give you what you want. When an experimenter asked passersby for a dime to make a telephone call, only 29 percent of them fished in their pockets for him. But when the request was accompanied by a brief touch on the arm, more than 50 percent complied. In another experiment, grocery store employees offered shoppers an opportunity to taste new products. Only when touched on the arm were shoppers more likely to stop and try what was on offer. Job applicants touched during an interview were more likely to agree to do volunteer work. Students touched in the classroom were more willing to go to the blackboard. Customers touched by waitresses ordered more alcohol and left bigger tips.

In this context, it comes as no surprise that a light touch may also make a person more agreeable in a dating context. Nicolas Guéguen, a social psychologist at the University of Bretagne-Sud in France, sent a cute twentysomething guy with the code name Antoine to a French nightclub for three weeks. Antoine was instructed to approach women and ask if they'd like to slow-dance

Have the Upper Hand

In your relationships, who has the upper hand? I mean literally. Among couples, there is often an imbalance between who touches and who gets touched, which may represent the balance of power in the relationship. Women, from the time we were little girls, are touched more often than are males, and men tend to have the upper hand. In a study of fifteen thousand heterosexual couples, researchers found that when lovers hold hands in public, the man's hand was most often the dominant one, on top of the woman's. Another study found that the woman was most often on the side of the man's dominant hand (on his right if he was right-handed and on his left if he was left-handed). Whether on a date or in a long-term relationship, be aware of who dominates when holding hands, touching, kissing, maintaining gaze, initiating sex, keeping the walking pace, and beginning or ending a moment of intimacy. That said, there are other age-old forms of power that are often the domain of women, such as withholding and resisting intimacy. Touché.

with him, and to lightly touch the forearms of half the women as he made his overture. Remarkably, of the women he touched, 65 percent agreed to dance with him, compared to only 43 percent of the women who consented without any physical contact.

Later on, in a second trial, the handsome Antoine was sent to approach young women on the street. His assignment was to tell each one he thought she was pretty and ask her for her phone number for a future date. Again, he touched half the women as he made his request. As startling and creepy as Antoine's approach sounds, more than 19 percent of the women he touched gave him their numbers, compared to only 10 percent of the women with whom no physical contact was made. A female experimenter then approached the women Antoine hit on and asked them questions about their impressions of him. It turned out that the women Antoine touched found him more sexually attractive, dominant, and strong.

Even the briefest touch is powerful because it's a cue of dominance and status, according to Guéguen's study. Men do it to women more than women do it to men, and it often works in men's favor because women generally prefer high-status, dominant men. Of course, touch changes with context. A simple touch may also convey friendliness, sincerity, and understanding. Culture affects the perception of touch—if you're from a low-contact culture, you might shun it. The same goes if you don't like the person who is touching you, or if you avoid intimacy. When women touch men, it's interpreted as friendliness. Due to men's overperception bias (page 164), even grazing the fingertips on a man's shoulder may be interpreted as sexual interest.

It makes sense that touch usually enhances a person's attractive qualities, yet it's something we rarely think about. Try it the next time you want to make a strong impression, but avoid heavy-handedness. As Voltaire put it, "To enjoy life, we must touch much of it lightly."

What's the hidden agenda in men's pickup lines?

A man sidles up to you at a bar. You can't make out exactly what he looks like in the dim light, but you can hear him . . . unfortunately. In a booming voice he shouts, "Hey there, I may not be Fred Flintstone, but I bet I can make your bed rock!"

The "Fred Flintstone" is one of the pickup lines showcased in two studies of male flirtation by British psychologists Matthew Cooper, Rory Morrison, Christopher Bale, and their colleagues. If you were to ask them for insight on why a man would ever use a line like the "Fred Flintstone," they'd tell you it's a sexual display. Why, you'd ask, would anyone intend to display bad taste? What's the hidden agenda here?

The answer is that whether or not you respond warmly to a man's pickup line tells him something about your personality and what you seek in a mate. The more controversial or sexual or humorous the come-on, the better it is as a filter. Consciously or not, a man uses these lines as a way to reject (or not waste time on) women who aren't looking for someone like him. A man using the "Fred Flintstone" line is more likely seeking a lay, not the love of his life, and any woman who doesn't want to club the Neanderthal must also be seeking sex. Like my friend Rita. At a bar, a man slunk up to her and asked, by way of introduction, if she'd ever had sex on top of a washing machine. She rolled her eyeballs and laughed. The dirty language didn't bother her. She actually thought he was funny and sexy, and they had a fabulous roll in the sheets.

The psychologists asked several hundred men and women to consider the variety of opening gambits that men use when picking up women, how successful they are in general, what type of guy tries each line, and what type of woman falls for him. Participants were asked to take a standardized personality test that measured their levels of psychoticism (rebelliousness or recklessness), extroversion (sociability), and neuroticism (emotionality, depression, and anxiety).

Overall, women responded best to pickup lines that reflected character, culture, resources, spontaneous wit, and ability to take control of a situation—qualities that represent "good mates" (introducing himself to a woman after defending her space in the bus line, telling her that her drink is on the house and then introducing himself as the club owner, or asking her to help him pick out an expensive watch). Humorous lines ("Haven't you forgotten me?" or "Can I buy you an island?") were rated as somewhat effective and filtered out introverted women, who appreciated them less than other personality types. Compliments ("I'm late and you're stunning" or "You remind me of a parking ticket because you've got fine written all over you") were not as successful on most women, and probably were perceived as too glib. However, women who scored high in neuroticism warmed up to these gushy lines more than other personality types did. Rated least effective by most women, but less off-putting to those who score high on psychoticism, were the sexually loaded come-ons such as the "Fred Flintstone" line and the no-nonsense "Do you want to have sex?" gambit. (The study's weakness is that it didn't involve real-life situations, whereby appearance and approach can make bad lines sound good, or vice versa.)

The fact that women usually go for "good mate" opening lines reflecting character, culture, wealth, and perhaps humor is in keeping with the theory that women generally prefer men who have potential as long-term mates. But women aren't always looking for the love of their lives; some deliberately seek sexy bad boys for casual relationships. That's why the "Fred Flintstone" come-on and those like it sometimes work. Call my friend psychotic, but she fell for the "sex on the washing machine" line because she wanted a sexy, rebellious man at that time in her life. It wasn't a strong foundation, nor was it intended to be. But it was great bed rock.

Why is blushing sexy?

Freud called the blush a displaced erection. You can't help it any more than an adolescent boy can help his hard-on. That spread of red is under the control of the sympathetic nervous system, which also makes your pupils dilate, your heart thud, your armpits sweat, and your limbs break out in goose bumps, among other involuntary acts. (Who's the sympathetic nervous system sympathetic to? Not you, it seems.)

When it comes to attraction, your blush, also called a "color signal," reveals something about your state of mind. The surge of blood in the delicate facial veins is a universal cue that means a person is aroused, amorous or ashamed. It says, coyly, *I'm smitten and self-conscious!* A woman's bright pink hue might be the clue some men need to know she's romantically interested. Blushing is like flashing your heart.

Blushing is also a way of advertising your youth. Your slow, burning blush blurts, *I'm experienced enough to know what's going on but innocent enough to be embarrassed or excited.* In a study at the University of California at Davis, 64 percent of people age twenty-five and younger reported blushing more than once a week (and 36 percent blush daily), compared to only 28 percent of people over twenty-five years old. (Think about it: when was the last time *you* blushed?)

A blush also suggests health and sexual arousal. If you can blush, your body is capable of dramatic physiological upheavals. Your heart is pumping more strongly than usual (thanks again to the sympathetic nervous system), and the blood that flows is full of oxygen-rich hemoglobin. Again, the older you get, the less color reaches your cheeks.

There's an interesting little evolutionary theory about blushing. According to neurobiologists at Caltech, blushing was so crucial to our primate ancestors, who depended on the color cue to perceive emotions and sexual signals, that it's linked to the development of color vision. Human eyes are optimized to pick up reddish wavelengths that correspond to subtle shifts in skin tone. Most people can detect a blush on faces of all complexions.

On another evolutionary note, blushing may also help explain why humans—even full-bearded men—have hairless cheeks. In some other primates, arousal is expressed elsewhere on the body. When a chimp is in heat, it's her butt cheeks that turn bright red.

Why does mimicry make you more likable?

Think back to a date that went well—how natural the interaction was, how in tune you felt. Only if you were very self-aware would you have caught yourself mimicking your date's accent, tone, pauses, even his catchphrases. You also might've been mirroring his body language—your elbows on your knees, with the same lip-biting little smile. And when he became optimistic and expansive, you probably felt very happy, too, and made the same long, sweeping gestures. If he felt the same way, it's likely that he was also mirroring you.

Tanya Chartrand and John Bargh, psychologists at Duke and Yale, respectively, led a particularly effective mimicry study. While sitting in a room having a conversation with a subject, an appointed experimenter subtly but deliberately mimicked the subject's posture. If the subject leaned forward, the experimenter also leaned forward, and if the subject leaned backward, so did the experimenter. This happened so naturally that the subject was oblivious that he or she was being mimicked in the course of the normal interaction. Later, when subjects were asked to rate the experiment, those who had been mimicked reported that they liked the experimenter and perceived the interaction as having been smooth, whereas participants who hadn't been mimicked gave lower ratings. In a similar study, subjects who were eager to get along with the experimenter in an effort to accomplish a task together unconsciously mimicked the experimenter's foot-shaking and face-touching gestures.

Psychologists observe that if you have a motive to bond with a person—for example, if you need to work together or you're sexually interested—you are more likely to mimic his or her expressions and body language. Mimicry happens in group situations,

Find a Partner Whose Moods Suit You

The more empathic you are, the more likely you are to detect your partner's moods and mirror them as part of the way you bond. In studies of depressed people in relationships, the unhappy half of the couple often becomes an unintentional "mood driver," railroading his or her partner into a similar yet more diffuse funk. Mood contagion is a common phenomenon; even mimicking another person's distressed facial expressions or vocal tone can induce the same state in you. (It also helps explain why you feel jittery after a horror movie or aggressive after watching a debate.) On the upside, you can also experience happiness vicariously. So picking a partner is, in a way, like picking a TV channel. Decide what you like, because that's where you'll be tuned.

too. A sort of smooth synchronicity develops, as it does in flocks of birds and schools of fish. From an evolutionary perspective, the ability to mimic was crucial in an environment where human survival depended on group cohesion. By unconsciously adopting the behaviors of others, our ancestors became more congenial and communicable.

By mirroring your date, you're not just charming him—you're also showing understanding, empathy, and connectedness on a neurological level. It's a virtuous cycle: liking leads to mimicry, and mimicry leads to liking. That's because mimicry often genuinely influences your emotions. It's why happy people perk you up and depressed people bring you down. Neuroscientists have found evidence that emotional mimicry is due to a bundle of neurons located in the right inferior frontal gyrus of the frontal lobe known as a mirror neuron system (MNS). The MNS, which has also been found in monkeys and other animals, fires up when you pick up on the behavior of others. Mirror neurons enable you to understand another person's actions, intentions, and emotions.

Using functional magnetic resonance imaging (fMRI), neuroscientists have found that when people imitate angry, sad, or happy faces, their mirror neurons essentially reflect the activity they see, as if they had a firsthand emotional experience. (Test yourself sometime and deliberately imitate another person's expressions; you might notice a shift in mood.) In fact, all some empathic people have to do is to imagine another person's emotional situation for their own brains to undergo similar patterns of activity. Mimicry may also involve facial feedback—the theory that facial expressions modulate our feelings due to a common pathway between the motor (movement) and emotional regions of the brain. Mirror neuron activity has been found to be stronger in women than in men, and there's evidence that mirror neurons in the brains of people with autism are less active.

Try to catch your own facial expressions when you feel

you're really connecting with a date. You're probably mimicking his expressions, which may in turn modulate your mood. Or maybe you're mimicking his mood, which in turn modulates your expressions. And if he's mimicking your expressions, this can modulate his mood, which in turn affects your mood and your expressions. After a while you see that everything is interconnected—your expressions and your feelings, and his.

Why do you turn your head to the right when you kiss?

When biopsychologist Onur Güntürkün spied on lovers at airports, railroad stations, parks, beaches, and other public places, he noticed a routine choreography. Most of the couples, as they drew close to kiss, would turn their heads to the right instead of the left. Wherever Güntürkün went—the United States, Germany, Turkey—it was the same pattern, and septuagenarians did it as often as teenagers. Out of 124 couples, nearly 65 percent swiveled and tilted their heads right before locking lips.

A few years later, a group of Irish psychologists in Belfast replicated Güntürkün's voyeuristic study. This time, of 125 kissing couples, 81 percent turned their heads to the right. Careful to eliminate the influence one kisser might have on another, the researchers asked 240 volunteers to kiss a doll's face. The pattern remained: 77.5 percent tilted their heads to the right before planting a kiss on the doll's mouth. Whether a person was left- or right-handed, the chances of him or her turning right to kiss was higher than turning left.

According to Güntürkün, the reason most people kiss to the right is that we have a right-side motor (movement) bias. In the womb, beginning at twelve weeks' gestation, most fetuses turn their heads to the right and also move their right arm more than the left. Because the right-side motor bias occurs so early on, researchers conclude that it's genetic and related to muscular and spinal development. The fetus's innate bias for right-side movement may play a role in subsequent neural development, causing or enhancing differences in perception and motor control between the left and right hemispheres of the brain (such as spatial awareness and handedness).

Babies tend to sleep with their heads turned to the right. Two out of every three of us favor our right foot, eye, and ear. Eight out of nine of us are right-handed. And we're not alone: fish tend to turn rightward when a predator faces them head-on, chicks turn their heads right as they hatch from their eggs, and rats turn right when trying to escape a maze (interestingly, rats with a left-turning bias tend to have weaker immune systems).

Emotions also appear to give us a slight right-turning bias. Expressions are expressed more intensely on the left side of the face, which means that when you turn your head to the right you expose more of your emotive left side. This is because the right hemisphere of the brain, which controls the left half of the face, is related to moods, while the left hemisphere of the brain, which controls the right side of the face, is more analytical.

Indeed, when researchers Michael Nicholls, Danielle Clode, and their colleagues at the University of Melbourne asked subjects to strike an emotional pose as in a family photo, both genders tended to turn their heads right more often than when asked to pose as a rational, self-controlled scientist (approximately 60 percent versus 40 percent). Similarly, in a study of 1,474 paintings, nearly 70 percent of the portraits of women, such as the Mona Lisa, and 56 percent of the portraits of men show subjects with their heads turned to the right, thus revealing more of the emotive left side of the face.

When Nicholls, Clode, and their colleagues studied the effect of right and left poses on perceived emotion, they concluded that faces turned just 15 degrees to the right, which show a sliver more of the left side of the face, reveal significantly more emotion than photos taken head-on or of faces turned left, even when the images are mirror-reversed. (However, the right side of the face is often perceived as more beautiful and powerful; see pages 10–11.) Showing the emotional left side of your face might influence your kissing behavior—maybe you're more likely to turn right when you're emotional and left when you're self-aware—although researchers suggest the motor bias is a stronger influence. After all, when smooching dolls, people turn their heads to the right almost as often as they do when smooching people.

Know Your Rights

Most of us have an unconscious right-side motor bias. It's useful to know this if you've ever been anxious about a first kiss—chances are at least two in three that he'll tilt his head right, so you should, too. (Interestingly, lefties are also predominantly right kissers.) Knowing that two-thirds of us have a right-side bias can be helpful in other ways, too. When walking into a theater or any other crowded room, chances are better that you'll find seats on the left side of the room because, as studies have shown, most people instinctively migrate right. (Savvy stores know this and stock their newest items to the right of the entrance.) Likewise, take the left exit in the subway or a stadium to avoid a crowd. The emotional aspect of the right-turning bias is also useful to know. According to the University of Melbourne researchers, when you're expressive and emotional, you might subconsciously turn your head to the right, showing more of your moody left side. Likewise, when you're in control or feeling self-aware, you might turn your head to the left, exposing your more attractive and "powerful" right side. Even a minor head rotation may dramatically affect the way people perceive you.

Why do we French-kiss?

On the face of it, French kissing doesn't seem so sexy. There you are, with your tongue down your partner's throat, swapping saliva and swarms of bacteria. Thirty-four of your facial muscles are fully engaged. Your lips are swollen from all the blood rushing through your body, thanks to your heart, which is pumping at up to twice its normal rate. Your pupils dilate, your groin tingles, and you feel feverish.

French kissing (also called deep kissing or soul kissing) might not sound good, but when it's right it sure feels good. According to a study of more than a thousand college students at the University at Albany, most women regard the first kiss as a make-it-or-break-it moment. Only 10 percent of women said they'd even dream about having sex with a guy without kissing first, whereas men didn't think such preliminaries were so important. Yet kissing is a compatibility test. Are your partner's lips supple and sensitive, or tight and tense? How does he hold you—hungrily, with passion and wonder? Or weakly and ceremoniously? Is he even the kissing kind?

Your lips are dense with sensory neurons. If you like your date's kiss, nerve endings shoot happy signals to your brain's cortex to release neurotransmitters, including dopamine and endorphins. Dopamine fuels your brain's reward system, which motivates you to continue kissing. Endorphins, also known as natural painkillers, enhance pleasure (the likely explanation for how a kiss can turn a frog into a prince). Levels of cortisol, a stress hormone, drop in couples who kiss. Smooching may also flood your brain with oxytocin, the "hug drug" hormone that strengthens your attachment to another person and makes you feel warm and cuddly (see page 144).

Evolutionary psychologists believe that kissing is part of a courtship ritual to judge a potential mate's body chemistry and compatibility, which is why at least 90 percent of human cultures do it. By getting close to your partner and even tasting his saliva, you capture his "chemical fingerprint." His saliva and sweat contain potential pheromones that either turn you on or turn you off. Do you like his taste, or, as a friend put it after a disastrous exploratory effort, did "something die right then and there"?

Kiss for Your Health

In addition to releasing "feel-good" neurotransmitters and hormones, deep kissing may soothe and strengthen your immune system. Hajime Kimata, a Japanese researcher, instructed nearly fifty patients with eczema and allergic rhinitis to kiss their lovers or spouses for a half hour straight while listening to soft music. Before and after the smooching sessions he measured the levels of each subject's immune system components in their blood. Kissing decreased the production of allergen-specific antibodies, therefore reducing the allergic reactions that lead to bad skin and allergy symptoms. By making out, it seems, you might need less makeup.

CHAPTER 7
Sex and Seduction

He looked into my eyes, pinning me like a
damselfly on corkboard.
He spoke in my ear using a low voice, not a whisper.
"Do not move. Spread your legs just a little," he said.
His lips barely brushed my ear.
He proceeded to tell me what he would do
to me, with me, on me.

—Kelly Clayton, from "What I Saw in Him"

Why do men have more casual sex?

In a classic sociology experiment, a pretty gal and a cute guy approached opposite-sex strangers on a college campus and casually, in the course of conversation, asked if they'd like to have sex. Seventy-five percent of the men approached said yes, and those who declined apologized profusely and offered excuses, such as that they were going to dinner with their fiancée's parents. But not a single woman accepted the man's generous offer. That's not to say women don't have casual sex. They do, of course—but a major difference between men and women is that women don't generally want as *many* short-term liaisons as men want (and sometimes have).

The International Sexuality Description Project, a survey of more than sixteen thousand people worldwide, confirmed a well-known fact: men, regardless of age, marital status, or sexual orientation, are universally more interested than women in having an abundance of lovers. In almost every corner of the planet, one of every four men desired more than one sex partner in the next month. Only one in every thirteen women sought the same, and that was in Eastern Europe, where women report having the most flings. Fewer than one in every thirty-three American women stated they wished to have sex with more than one man monthly (except perhaps in their fantasies). And among the horny people worldwide who admit they are actively seeking casual sex partners, only 20 percent of the women are looking for more than one lover, versus 50 percent of men (although they probably won't get them).

Even if you're a women looking for a short-term relationship, there's statistically little chance you'd sleep with someone you just met. In a study involving 150 U.S. college students, the average man declared that he'd have sex within one week of meeting a woman, whereas women were less inclined until several months had passed. (In the large cross-cultural study, however, women from Western Europe, Australia, and New Zealand were much more likely to have sex with a guy within a few weeks of meeting him.) How much men value chastity in a mate depends on culture. Men in Africa, the Middle East, and South America generally value chastity the most, while men in Western Europe, Australia, and New Zealand value it the least, indicating that prior sexual experience is irrelevant in a mate. In fact, lack of experience might even be objectionable. Most cultures fall somewhere in between.

From an evolutionary perspective, women have very little incentive to have loads of lovers. A man can have sex with a hundred women in a year and have a hundred babies. His genes can be spread far and wide, and from that perspective he has an incentive to be a playboy. If a woman has sex with a hundred men, she'll only have one child at the end of her yearlong lovefest (unless she has twins). There's no reproductive benefit for women to have so many lovers; in fact, there's a disincentive. Any one of those guys could turn out to be the father of the child that you must carry in

Where to Have a Fling

Culture influences people's openness to casual sex, although everywhere around the world far fewer women are into it than men. If you're wondering where on the planet people are most promiscuous, look no further than the International Sexuality Description Project's (ISDP) survey of more than six thousand young people around the globe. Among other juicy factoids, it reveals that the percentage of men yearning for more than one sex partner is highest in South America (35.0 percent), the Middle East (33.1 percent), and Southeast Asia (32.4 percent). For women, it's Eastern Europe (7.1 percent), Southeast Asia (6.4 percent), and South America (6.0 percent). The only regions in the world where men said they'd be *unlikely* to consent to sex after knowing someone for one month are East Asia (due to conservative sexual attitudes that the researchers link to the male-biased sex ratio) and Africa (due to the AIDS epidemic; rural and uneducated populations are underrepresented). Men who are the most likely to immediately jump in the sack with strangers hail from Oceania (Australia and New Zealand), South America, the Middle East, and Western Europe. The only region where women report a slight likelihood of having sex with someone they've known less than a month is in Western

Europe, although Oceania is a close second. As noted by the ISDP, differences between the sexes remain constant throughout the world, yet there are clearly cultural factors in women's attitudes toward casual sex, including a more liberal sexual outlook, a higher ratio of women to men, weaker religious ties, and, in the case of Western Europe and Oceania, relatively low fertility rates and gender-egalitarian political systems.

your body for nine months, then nurture until adulthood and beyond. The better strategy, at least in the days before birth control, was to be more selective about your lovers and therefore have fewer of them.

Men and women alike agree that the number one reason for having a short-term relationship is raw physical attraction. When women have casual sex, they tend to go for guys who are socially or physically dominant (a boss or college professor) or good-looking (the cute guy in the next cubicle or the jock at the gym). Women prefer tall, well-built, symmetrical men, and men home in on women's faces, hair, and bodies. When it comes to flings, men just about end their requirements there, but women often seek non-superficial qualities as well: creative, mysterious, gallant, well-rounded, fun, charming, warm, memorable, and so on. This might have a lot to do with the fact that women and men differ in their second reason for having casual sex. Women are motivated by the possibility that the fling might lead to something more serious. For men, reason number two is that the fling gives them a sense of their "mate value," that is, their desirability compared to other men (see page 260).

Birth control has no doubt reduced the number of kids born of passionate love affairs. Ironically, oral contraceptives—the celebrated driver of the sexual revolution—may also reduce the hormonal influences that draw women to the type of men with whom they'd have casual sex (see pages 34 and 120).

Why are fewer men than women bisexual?

Men's sexual orientation is usually binary, gay, or straight, with few guys identifying themselves as bisexual. That's what psychologist Richard Lippa at California State University found when he analyzed data submitted by more than two hundred thousand people of all ages worldwide. It's the straight woman who isn't so straightforward. She's *twenty-seven times* more likely than a man to have a homosexual experience. The higher a woman's sex drive, the more likely she is to have many lovers and be actively bi. The higher a man's sex drive, the more likely he is to have many lovers of one gender only.

Why are there so few bisexual guys? To find out, psychologists Meredith Chivers, Gerulf Rieger, and Michael Bailey at Northwestern University enrolled a hundred heterosexual men and women and bisexual men in their study on sexual arousal patterns. Sent to private rooms, the women were outfitted with tampon-like devices to measure vaginal blood flow, and the men were collared with instruments that recorded changes in their penis girth. Then they were instructed to lie back and watch pornographic videos. The porn was representative of two genres: straight (only women-on-women was shown; results could be thrown off by the presence of a stud) and gay (men-on-men).

The upshot? Straight men were more sexually aroused by the straight porn. Gay men were more aroused by the gay porn. Straight women got hot and bothered by both male and female nudity, supporting the evidence that women's sexuality is fairly fluid and bisexuality is common even among women who call themselves straight. Bisexual men, however, had completely unexpected results. While you'd think that bi guys would be turned on by straight *and* gay porn, 75 percent of them had arousal patterns that matched gay men's, even when they said they were subjectively aroused by both genres. The other 25 percent had arousal patterns similar to straight men's.

The researchers conducted a subsequent study on more than a hundred men of all sexual orientations and reached the same conclusion. Although bisexual men have the capacity to be slightly aroused by women, most are turned on more by men.

Research of this ilk has since become controversial for casting doubts on whether male bisexuality exists. But just because bisexual men aren't physically aroused by men and women equally doesn't mean they don't have a legitimate sexual orientation.

According to some clinical psychologists, guys who call themselves bi may actually be attracted to both sexes but split in two directions: to men physically and to women emotionally and romantically (or, less commonly, the other way around). An alternative theory, difficult to measure by genital arousal to porn, is that bisexual men are attracted to the mental or psychological rather than physical gender of a person (i.e., feminine-looking and -acting people of either sex).

Sexual orientation is determined by genetics, brain anatomy, exposure to prenatal sex hormones, and cultural factors. Future studies on the brain, hormones, and psyche may yield more insights into the middle way of bisexuality.

Are men more aroused than women by pornography?

Yes, sexy visuals have a greater impact on men than on women, but it's a misconception that pornography (or erotica) is only a guy thing. By and large (and XXX large), women enjoy it, too. In fact, in studies that measure blood flow to the genitals, porn-watching women are *more* physically aroused than porn-watching men. Women reach the same peak genital and thigh temperature as men do, in the same amount of time, and report similar subjective

Try Female-Friendly Porn

Why not find an erotic flick of your choosing and watch it with your partner? One reason why some women aren't into porn is that so much of it involves incessant anonymous sex, which just doesn't seem so appealing. (In a large British survey, only 15 percent of women fantasized about having sex with two or more men simultaneously.) In evolutionary terms, women benefit little (and risk a lot) reproductively from having excessive sex with strangers, whereas men have little risk and lots of upside (literally). There are plenty of sexy videos that aren't of the male-oriented *Playboy* variety and focus on gratification for women as well as men. Check out erotic films by female directors such as Petra Joy, Candida Royalle, Anna Span, and Nelly Kaplan.

arousal levels. (Using thermal imaging devices, researchers have found that it takes about ten minutes for both men and women to reach peak arousal.) Of course, it's hairy to say that *all* women are enthusiastic about porn. Culture is a big factor, and so is content and context. You might enjoy a porn flick only when the vigorously fornicating characters are in loving, equal relationships, not a misogynistic rut.

To measure how porn affects the brain, neuroscientist Stephen Hamann and his colleagues at Emory University took functional magnetic resonance imaging (fMRI) scans of straight men and women as they gaped at erotic photos of naked hotties and copulating couples (as well as neutral control stimuli). Both sexes said they felt turned on when looking at the porn, especially when looking at the couples. Indeed, everyone's brains showed activation in certain areas: the anterior cingulate, which is linked to attention, emotion, and sexual motivation; and the reward areas of the nucleus accumbens and ventral striatum. But the similarities ended there, because men's brain activity in two regions, the hypothalamus and the almond-shaped region known as the amygdala, was strikingly different from women's. During the viewing of erotic imagery, these "appetite centers" of men's brains showed significantly increased activity, while women's were relatively quiet.

The hypothalamus initiates the release of sex hormones and fuels erotic feelings and reactions such as a faster heartbeat. It responds to the amygdala, which regulates emotional arousal, motivation, and spur-of-the-moment decisions. If your amygdala is hyperstimulated, you react strongly and impulsively. The amygdala is also strongly linked to the sex drive—the larger your amygdala, the more powerful your drive. (Imagine: if we all had fMRI vision, we could scan a room for people with high sex drives by looking at their amygdalae.) Men have larger amygdalae than women do, proportionally, which is one reason why their sex drives are often stronger. According to the researchers, the activation of the amygdala when watching porn (and its effect on the hypothalamus, which secretes sex hormones) helps explain why men have a greater appetite for it and a greater urge to act on or interact with what they see. Once a guy reaches orgasm, his amygdala shuts down, and the erotic video or magazine is no longer so compelling.

It is not yet clear if the difference in how men and women process sexy images is because male and female brains are inherently dissimilar or because we just have dissimilar experiences. From an evolutionary perspective, it makes sense that the male amygdala responds so strongly to the sight of sex. After all, for countless generations, our male ancestors preferred mates based on visuals such as age, looks, reproductive status, and cues of sexual availability. The sight of sexually receptive women was fairly rare until mass media, and evolution favored men who seized sexual opportunities when they presented themselves.

The bottom line is that even if you and your partner like porn equally, he might be more motivated by it, react to it more strongly and impulsively, and seek it out more often. You might not see it the same way. Whatever tickles his amygdala and hypothalamus might only tickle your fancy.

Can a romantic movie set the mood for love?

Yes, movies can set a mood, sometimes for hours, and one way they do so is by affecting your hormone levels. Emotions affect hormones, and hormones affect emotions.

To find out how much movies move your hormones, psychologist Oliver Schultheiss and his colleagues at the University of Michigan asked men and women to watch thirty-minute scenes from three flicks: *The Bridges of Madison County* (a romance about a married woman falling in love with a handsome stranger), *The Godfather Part II* (a thriller in which a mob boss sadistically kills off his opponents), or a neutral documentary about the Amazon rain forest.

The psychologists were especially interested in how the different films would affect levels of two hormones, testosterone and progesterone. Testosterone levels are associated with social dominance and aggression. Progesterone is associated with bonding and anxiety reduction, and possibly a reduction in libido. The researchers assigned each of the sixty participants one of the three movies to watch. Their testosterone and progesterone levels were measured immediately before and after the viewing, and then measured again forty-five minutes later. Before and after, the subjects also took a standard psychological test to determine their motivational state, including their power motive (need for dominance, associated with testosterone release) and affiliation motive (need to bond and connect, related to progesterone release). The movies were intended to induce these motivational states and measure their impact on hormone levels.

From a dating perspective, the results show that you can choose a movie to set a mood. But first decide what type of night you want.

If you want to raise a man's testosterone, go for something like

The Godfather series. Within forty-five minutes after watching gory and violent scenes from those movies, men's testosterone levels surged by as much as 30 percent if they had high baseline levels of the hormone. The study didn't measure aggression or personality, but it appears that violent thrillers could make full-of-fight types even more wound up. Men with a high power motive clearly identified with the mafioso's drive for domination, which triggered their hormones. This is probably great news if you're into aggressive, dominant men—maybe it'll fire him up and you'll have great sex that night. But it's probably not-so-great news if you're looking for a little romance after the credits roll. Other studies show that powerful speeches and sporting events have the same testosterone-boosting effect.

Among men and women whose baseline testosterone levels were low, watching the *Godfather* films didn't have much of an effect. Oddly, if women had a high baseline, their testosterone levels actually dropped, even though they showed a relative increase in power motivation. The reason for the drop is unclear. Perhaps watching two power-hungry men duke it out is threatening to women who otherwise strive for control, or not arousing enough. Further studies using films that depict women vying for dominance would provide greater insights. (The rain forest documentary had no hormonal effect, but maybe it would if you and your date are radical environmentalists.)

If you want to snuggle, try a film such as *The Bridges of Madison County*. Within forty-five minutes after watching Meryl Streep and Clint Eastwood make love in that movie, the testosterone levels of high-testosterone men tumbled significantly (although the number of high-testosterone guys watching the film was too small to average). According to the researchers, the plummet makes sense. Men's testosterone levels also tend to drop when they fall in love (see page 148). Moreover, men and women tested for increased affiliation motive after watching *Bridges* had a 10 percent increase in progesterone levels. Because progesterone potentially has bonding and calming effects, this could be a good thing for romance.

In light of this study, choosing the loving, generous, gushy, sexy, weepy film for your dates seems like a good idea. Insist on it, even if the guy prefers action thrillers and violent dramas. He may squawk about watching a "chick flick," but it'll make him more lovey-dovey than a hawkish thriller.

Are good dancers also good in bed?

"Nothing is more revealing than dance," quipped Martha Graham. Little did she know that by saying this she was in lockstep with Charles Darwin, who considered dance, with all its twists and turns and pelvic thrusts, to be a person's way of revealing his or her "good genes" to potential mates.

To study the relationship between dance and the body,

Make a Dance Date

Make him shake his booty. The more researchers look into the physical qualities associated with good dancing, the more they realize that Darwin was right: dance is a way to display (or judge) sexy genes. Guys with the best moves tend to be the ones with the most symmetrical bods. While women are better than men at detecting the best dancers, symmetrical men are better than average at perceiving talent in women's dance moves (so brush up on your tango to catch the eye of a hot guy).

Unsure of your dance skills? According to anthropologist Lee Cronk, who coauthored the study, dancing is so widespread as a form of courtship that it's best to work with it rather than shy away. Just as makeup may cover a poor complexion, dance lessons may conceal bad coordination. Oh, and one more tip: you may be attracted to a man who moves like Fred Astaire or Mikhail Baryshnikov when he's dancing freeform. But before you test the "good on the dance floor, good in bed" theory, make sure he dances just as well with you as he does alone.

Rutgers anthropologists went to Jamaica, where dance clubs are a big part of the dating scene, and videotaped nearly two hundred teens as they grooved to a pop song. Researchers Lee Cronk and

William Brown and their colleagues disguised the appearance and identity of the dancers by using motion-capture technology to create computer animations, and then asked the dancers' peers to judge each performance. The researchers also measured nearly every part of the dancers' bodies that come in twos: arms, legs, elbows, wrists, knees, ankles, feet, fingers, and ears. Some of the dancers were very symmetrical, with nearly matching left and right sides, while others were more lopsided.

Darwin wouldn't have been shocked to learn that the most symmetrical men and women turned out to be the best dancers. Anatomically it makes sense: symmetrical people may be more coordinated and have other qualities that shine on the dance floor. Among male dancers, symmetry accounted for nearly 50 percent of the variance in dance ability, whereas in female dancers it only accounted for 28 percent (experience and lessons may compensate). As judges, women were the pickiest and most perceptive. Regardless of their own ability, they were pros at picking up on a man's coordination, balance, timing, and rhythm. Men weren't as perceptive of women's talent unless they themselves were symmetrical.

Symmetry is a measure of genetic quality. Similarity between the left and right halves of your body is a cue of developmental stability—a body that's grown in the absence of parasites, toxins, disease, or genetic mutations. It makes sense that women are better than men at perceiving symmetry. We look for signs of "good genes" in our sexual partners, and dance reveals something about a man's overall developmental fitness.

So how might a dancer's skill at the bump and grind swing

over to the bedroom? It turns out that symmetry, the innate physical quality of good dancers, is also a predictor of sexual ability. In a study of symmetry and sex appeal, evolutionary biologists Randy Thornhill and Steven Gangestad at the University of New Mexico recruited nearly seventy-five men and, just as the anthropologists did with dancers in Jamaica, measured the length and breadth of the men's body parts and calculated the differences between the left and right sides of their bodies. Then they brought in women to rate the men's attractiveness. Overwhelmingly, and without knowing the measurements, gals preferred guys who were the most symmetrical. The hottest guys had asymmetries around 1–2 percent per body part, whereas the homeliest guys averaged a 5–7 percent difference per body part.

As a group, symmetrical guys also turn out to be the most sexually experienced. In a study of 122 men and women, Thornhill and Gangestad found that men with symmetrical bodies tended to have sex three to four years earlier and have two to three times as many sexual partners in their lifetimes than men with more irregular bodies. (They're also bigger cheaters, citing more affairs.)

The clincher is Thornhill and Gangestad's subsequent study on symmetry and sexual talent. Of nearly ninety couples, women with the most symmetrical boyfriends claimed to have orgasms 75 percent of the time on average and had a tendency to climax simultaneously with their partners. In contrast, women with the most lopsided lovers reported orgasming only 30 percent of the time. (As you read this, you're probably wondering where to buy vernier calipers.) Love had nothing to do with how turned on the women were by their symmetrical lovers. Nor did duration of relationship, degree of intimacy, sexual experience, or use of contraception.

Of course, scientists weren't the first to observe a special "symmetry" between dance, hot bodies, and sexual selection, but now there's more proof. This isn't to say that all great dancers are also great lovers—but their masterly control of rhythm, coordination, and timing can't hurt. They're in touch with their own bodies, and that bodes well for being in touch with yours.

Is chocolate really an aphrodisiac?

If chocolate has libido-enhancing chemical properties, they're limited and vary from person to person. Blame the Aztecs for planting the idea that chocolate is an aphrodisiac, and the Europeans for helping it take root. Legend has it that the cocoa harvest coincided with an Aztec festival of wild orgies, and by association the bean became a symbol of fertility.

Chocolate contains two neurotransmitters that might have a minor effect on the brain, tryptophan and phenylethylamine. Tryptophan, an amino acid also found in turkey, is famously known for making you feel content and drowsy. Phenylethylamine is a stimulant that produces the same sort of euphoric rush you feel when you fall in love. Unfortunately, chocolate contains only minute amounts of these chemicals, and phenylethylamine is rapidly metabolized in your body, meaning that very little of it actually makes it to the brain (much less the genitals). Chocolate also contains theobromine, a stimulant that can get your heart racing as much as caffeine does, especially if you eat a lot of dark chocolate. As heart-pounding as high-caffeine chocolate can be, studies reveal that there's no difference in sexual arousal between women who eat one to three servings of chocolate daily and those who eat none at all.

The more likely reason for chocolate's sexy reputation is its magnificent physical properties. While oysters, bananas, halved melons, and avocados all resemble genitalia, chocolate's sensuous appeal is its mouthfeel. It's rich, complex, and creamy, and it melts at body temperature. The addition of sugar and fat in chocolate bars and candy can raise your serotonin and dopamine levels, which might lift your mood if not your libido. Dark chocolate also contains flavonols, which are said to increase blood flow and prevent clotting, and thus help prevent strokes. Perhaps it's more helpful to think of chocolate in these terms: whatever's good for your heart can't be bad for your sex life.

How does alcohol affect your sex life?

As writer Kinky Friedman put it, "Beauty is in the eye of the beer holder." The more you imbibe, the more likely you are to have sex with someone you wouldn't even kiss, perhaps not even touch, when stone sober. This is in part because alcohol has an effect on cognitive capacity. When you drink, the reward areas of the brain that mediate sexual attraction—the nucleus accumbens and ventral striatum—become superstimulated with dopamine, making you feel good and making others look good. Alcohol also increases the release of endorphins, which block pain, and increases the efficiency of serotonin receptors, which makes you happier for as long as the alcohol is in your system.

Alcohol also affects testosterone levels, which does funny things to people's sex drives. In women, even a little tipple can cause a rise in testosterone levels and make some of us more sexually aggressive. In men, too much alcohol makes testosterone levels *drop* by suppressing testosterone production in the testes. Sure, drunk men tell women how much they want them and they seem aroused, but it's only due to a surge in luteinizing hormone, a secondary hormone that triggers testosterone production when levels are low. Try as he might, a drunkard may not be able to keep an erection. Chronic alcoholics may develop "man boobs," shriveled testes, bigger hips, and a plump, smooth chin due to a prolonged lack of testosterone. For some men, the damage is permanent.

While your date's libido flattens when he's blitzed, yours escalates, and may reach unexpectedly high levels depending on where you are in your menstrual cycle. In a study on alcohol consumption and hormone levels, Finnish psychologist Ralf Lindman and his colleagues asked women to keep a diary of their sexual feelings and activities every day for a month, as well as the number of drinks they knocked back. It turned out that gals felt horniest and had the most sex when they drank alcohol in the follicular and fertile phases of their cycles (right before and during ovulation). At this time in the cycle testosterone levels are already high, and alcohol sends them soaring.

The double shot of testosterone you experience when you drink throughout your fertile window might make you feel easily aroused and sexually confident. It's not a bad thing, unless you find yourself doing things you don't intend to do. The lesson here is to be aware of where you are in your cycle before you go on a drinking binge. When your beer goggles are rose-colored, the morning after might not be a pretty sight.

Can semen make you happier?

Semen has a very basic recipe: a few sugars, amino acids, cholesterol, enzymes, proteins, mucus, and, unforgettably, a dash of spermatozoa. It also contains hormones, including testosterone, estradiol, opioids, oxytocin, prolactin, and prostaglandins—all of which have potential feel-good effects. When you have unprotected sex, these hormones are absorbed through the walls of your vagina. Within an hour or two they show up in your bloodstream, and possibly your brain. Even if some hormones don't directly hit your noggin, their effects on the rest of the body can indirectly make an impact.

"Semen will make you less depressed" sounds like the most self-serving line a man could utter. Yet there's some truth to it, according to a study by Gordon Gallup Jr. and his colleagues at the State University of New York at Albany. The psychologists asked three hundred women about their sex lives: how often they had sex, how long since their last rendezvous, how long they had been in the relationship, and what type of birth control they used, if any. The subjects also took a standard depression inventory test that measures emotional well-being.

It turned out that, as a group, the women who went "bareback" scored as less depressed than the women who used condoms, regardless of the length of the relationship, frequency of sex, or the emotional bond they had with their partners. Sure, the condom users enjoyed sex, including the satisfaction that comes from intimacy and orgasm, but they didn't appear to benefit from the sustained antidepressant benefit experienced by the women who rejected rubbers. In fact, regular condom users were more likely to report having attempted suicide than women who never or only sometimes used barrier protection. On the depression test, assiduous condom users scored about the same as those women who were having lesbian-only sex or were abstinent.

Equally fascinating, only women who never, or only sometimes, used condoms got more depressed when they stopped having sex. The more days that had passed since their last sexual encounter, the more depressive symptoms they reported. In stark

Can the Condom (but Be Safe)

Hormones in semen that are absorbed through vaginal tissues are likely to be just as easily absorbed by the tissues of the mouth and rectum. That's why, according to Gordon Gallup Jr., oral and anal sex may also yield the same antidepressant effect as vaginal sex, although studies have not been done.

That sperm may have an antidepressant effect is an uplifting idea, but it doesn't warrant unprotected sex. Unwanted pregnancies and sexually transmitted diseases (STDs) will dramatically offset any happy effects of hormones. But if you're comfortable with your partner, and you're both monogamous, committed, and free of STDs, you might want to give it a try—for scientific purposes, of course. Contraception may be an issue. Researchers have yet to determine whether the hormones in semen will have the same effect on you when you're under the influence of hormones in oral contraceptives. After all, studies indicate that oral contraceptives affect women's sensitivity to smell, attraction to male faces and body odors, and the beautifying benefits of hormones related to ovulation (see page 34). Men who have had vasectomies have much lower levels of testosterone in their sperm, which may also diminish the effect.

contrast, women who used condoms most or all of the time did not report more depression if they went without sex for the same amount of time. According to Gallup, the women who were accustomed to unprotected sex were going through semen withdrawal. From an evolutionary perspective, a semen addiction would make sense: if there's something in it that makes you feel good, you'll come back for more. The hormones in semen—testosterone, estrogen, and/or others—that make it a potential antidepressant may also help couples bond, promote pregnancy, and make sex more rewarding for the woman.

Is the study all wrapped-up and airtight? Not entirely. Skeptics wonder if women who don't use condoms might have happier and more volatile personalities than their more cautious peers (the study did not establish each woman's psychological profile with and without semen exposure). Although the researchers measured and controlled for relationship strength and length and personality, some of these variables may be difficult to gauge.

The upshot is that if you're less depressed after having unprotected sex with a man, it may be the hormones in his semen, or it could just be your fancy-free personality. Only one thing is certain: if further research is pursued, it won't be difficult to recruit male volunteers.

Why do women have orgasms?

If you want to watch scientists get passionate, bring up the debate over the female orgasm. While the reason guys climax is obvious, there's no clear reason why women do. Everyone knows a woman can get pregnant even from unsatisfying sex.

One camp of scientists concludes that the female orgasm is nothing more than an evolutionary by-product. In the first couple of months after conception, male and female embryos have the same basic anatomy. Later on, a tide of testosterone washes over the male fetus; he grows a penis where the girl has a clitoris, and neural pathways develop in the genital area that will make his orgasm—ejaculation—possible. In essence, boys get the whole shooting works while girls just keep the trigger, and the female orgasm is a happy accident.

But other anatomists and biologists believe that nature always has a purpose. Some say the female orgasm helps forge a bond between couples. Perhaps orgasms help a woman choose good mates, leading her to fancy men who are fit enough to be good in bed and thoughtful enough to ensure she's satisfied. Or perhaps orgasm is a motivation to have sex. Zoologists have long known that female monkeys and apes experience pleasure that resembles women's orgasms.

The most compelling argument for the female orgasm is that it's a woman's subconscious way of influencing when and with whom she'll get pregnant. It harks back to the 1960s, when a woman had sex with a sailor and the condom got stuck up inside her cervical canal. The gynecologist who extracted it speculated that the lady's uterine and vaginal contractions must have been so strong that they sucked the condom right off the seaman. From this the "upsuck" theory was born: the female orgasm sucks semen up into the cervix like a vacuum, which increases a woman's chances of getting pregnant.

To test the upsuck theory, British biologists Robin Baker and Mark Bellis asked women to track the timing of their vaginal orgasms during sex and to collect the "flowback" of sperm that leaked out afterward. Analyzing the legacies of more than three hundred sex acts, they found that if a woman climaxes around the same time as her partner, or up to forty-five minutes afterward, she retains significantly more sperm than she would if she didn't come. The more semen that stays inside, the better the chance that one of the sperm will make it to the egg.

Inspired by these data, Baker and Bellis proposed that the female orgasm is a covert, cloak-and-dagger, and completely unconscious way that women select the sperm of men with better genes they can pass on to their children. It's like a war of the womb, where the gentleman who wins is the one who can bring the lady to orgasm. One study at the University of Texas at Austin found that the frequency of a woman's orgasms with her partner is a predicator of her desire to have a baby with him (after controlling for marital happiness, foreplay, and other factors). Women who wanted to get pregnant also reported greater participation in the sex act—by being active, a woman can synchronize her orgasm with her partner's, thereby increasing the chances of conception. Not that she does it with the conscious intention to get pregnant.

It also happens that a woman is more likely to have an orgasm with men who have the "good genes" cue of symmetry. According to a study at the University of New Mexico, women reach climax a mind-blowing 75 percent of the time with very symmetrical men, versus only 30 percent of the time with very asymmetrical men. Notably, a woman's love for her sex partner had nothing to do with the number of orgasms she had with him.

Are orgasms genetic?

Nearly one in three women report that they never or only infrequently climax, whether alone or with a partner. Some women aren't able to reach orgasm at all, ever. At least some of the blame lies in the genes. Your genes may help push you over the brink or pull you back.

Exactly how influential genes are appears to depend on context. A team of behavioral geneticists at the University of Chicago, led by Khytam Dawood, asked more than three thousand female twins how often they achieved orgasm in different situations. Each pair of sisters grew up in the same household and had the same upbringing. Identical twins, who share 100 percent of their genes, reported more similar orgasmic experiences than did fraternal twins, who share only 50 percent of their genes. The difference in results between identical twins and nonidentical twins reveals a lot about the contribution of genes versus environment. Comparing the data from the two twin sets, the researchers determined that genes account for 31 percent of a woman's success in climaxing during intercourse, 37 percent when having nonpenetrative sex, such as oral sex, and 51 percent when masturbating.

However, exactly which genes are involved in orgasm is a complete mystery, not to mention that climaxing during masturbation probably involves different genes from climaxing with a partner. Genes have different functions in different contexts. "Orgasm genes" may be related to your reproduction system, nervous system, limbic (emotional) system, endocrine (hormonal) system, or all of the above, interacting with one another and the environment in a complex way. Genes that may affect your personality, such as whether you're extroverted or confident, excitable or calm, could have a role, too.

Genes that influence the anatomy of your genitals and nervous system may play a particularly crucial role in your ability to reach orgasm. The fabled "G-spot" is an erogenous zone in the vagina that, when stimulated, leads to a powerful vaginal orgasm without clitoral involvement. There are anatomical differences between women who are able to have vaginal orgasms, an estimated 30 percent, and the majority who can't climax without at least a little external friction. Using vaginal ultrasound, researchers at the University of L'Aquila in Italy have found that women with G-spots have thicker urethrovaginal tissue, which is rich in blood vessels, muscle fibers, and glands. Add these anatomical differences together with other genetic influences and psychosocial factors such as personal well-being, education, and religion—and you realize that orgasms have mind-blowing complexity.

Why women are genetically different from one another in their ability to orgasm may have an evolutionary explanation. Orgasms may help women retain sperm, which in turn increases the chances of pregnancy. It's possible that genes that make the orgasm threshold a little higher give some women an advantage because they're more likely to climax only with men they find sexually attractive, thus increasing the odds of bearing children with these desirable qualities. On the other hand, genes that enhance sexual responsiveness may encourage women to have sex more often and increase the number of children they bear.

The whys and hows are still unknown. While reaching the big O may be a stretch for some women, it's even more elusive for scientists.

Test Your G-Spot

Italian researchers have found that women who can experience "no-clit" vaginal orgasms have thicker walls in the front of their vaginas than those who depend on external friction. Women who have vaginal orgasms describe them as deep and heaving in contrast to the shallower orgasm from clitoral stimulation. The magical erogenous zone, termed the *G-spot* after German gynecologist Ernst Gräfenberg, is relatively uncommon; researchers believe that only one out of every three women has it. Short of getting a vaginal ultrasound, you can test if you have a G-spot by massaging and applying pressure two to three inches up inside the vagina on the front wall (behind the pubic bone). It may be easier if you train your partner to strike that area during sex or rub his fingers over the area in a beckoning "come here" motion. Orgasm from G-spot stimulation requires heavy pressure. If you can reach climax from the G-spot alone, lucky genes might be the reason. Some researchers also suggest that the G-spot is like a muscle—with proper exercise, it gets stronger. For men, the erotic equivalent of the G-spot is the prostate gland, a bulb of tissue an inch into the rectum, which can be stimulated in the same way as the G-spot.

Do women really reach their sexual peak in their thirties?

Many women believe a libidinous tidal wave will rush over them in their thirties, but the theory doesn't quite hold water. It was popularized by the sex researcher/crusader Alfred Kinsey. Conducting national surveys throughout post–World War II America, Kinsey found that women ages thirty to thirty-four report more orgasms than do women of any other age. Yet there's no conclusive evidence that orgasm frequency translates into higher sex drive. Contemporary psychologists point out that orgasms are actually only one measure of a sexual peak. There are others as well, including sexual desire (i.e., "lust motivation"), which drives a person to have more sex.

To see if there's a link between age and sexual desire, psychologist David Schmitt and his colleagues at Bradley University surveyed approximately eight hundred women of all ages. Among other questions, they asked the women to rate their sexuality on scales of lust, seductiveness, abstinence, promiscuity, and infidelity. Compared to other age groups, the (relatively small number of) women who were between thirty and thirty-four reported a barely significant increase in lust, seductiveness, and sexual activity.

It's difficult to come up with a biological reason why women would have an early-thirties blip in sex drive. In Schmitt's study, women of all ages reported the same levels of promiscuity and infidelity. Women are not having more extramarital sex after they have children, nor are they more likely to have casual sex as they grow older. There is no evidence of an increase in libido-stirring hormones such as testosterone as we age. If Mother Nature does have a reason for an early-thirties peak, the researchers conclude, it's that it increases reproduction and bonding with your partner before the risk of conceiving a baby with genetic disorders sharply increases in your late thirties and forties. Alternatively, the fabled early-thirties peak may be unrelated to women's sexual desire, and relative only to the decline in men's sex drives (which really does occur as men hit their thirties and forties and testosterone levels naturally decrease).

While no one denies that hormones play a role in libido and estrogen levels decline after menopause, it appears that the sex drive for women is also strongly influenced by emotions and context (which in turn can influence hormones). The belief that you're *expected* to feel a peak at this age may drive your expectations. According to psychologist Roy Baumeister, women's sexuality is malleable, responding as much to culture and circumstance as to hormonal drive, if not more. Women, he reports, experience more variation in their sex drives than do men; religion, education, politics, and peers affect women's sexual behaviors more than men's; and women are less consistent about their sexual attitudes.

When it comes to sex, women simply seem to care more about meaning and context, which has a dramatic impact on libido. This means that an early-thirties peak in women's sexual desire and activity may be a combination of cultural expectation and an increase in comfort with their bodies, sex, and relationships. If this is so, there's no need to fear that your sex drive slips down a gear by the end of your thirties. Why not set your expectations higher and make your peak last longer?

Why is intercourse more satisfying than masturbation?

"Don't knock masturbation," said Woody Allen. "It's sex with someone I love." Poor Woody. Despite his defense of masturbation, there's hormonal proof that sex with another person really does do more for you.

The hormone is called prolactin. The more prolactin in your bloodstream when you climax, the greater the feeling of satiety after sex. Prolactin's job is to offset the effects of dopamine, a hormone related to pleasure. During intercourse your dopamine levels are higher than when you're alone masturbating because you're more turned on and there's more going on physiologically and emotionally. When dopamine takes a nosedive immediately after orgasm, prolactin loses its brake and soars higher than it would otherwise. It's like water building up behind a dam; once the dam is released, the water gushes more forcefully. Your prolactin levels after orgasm indicate how high dopamine levels were before orgasm.

Theorizing that there's something special about intercourse orgasms, medical psychologist Stuart Brody asked nearly forty heterosexual men and women to reach climax in a laboratory setting. Half of the subjects were instructed to have sex with their partners—on their backs, with their partners active on top of them and an intravenous cannula collecting blood from their veins. The other half were sent to private rooms to masturbate alone to a film. On another day, instead of sex, the same people participated in a control condition in which they watched a neutral film and did not climax at all. When everyone's blood was tested, it turned out that the subjects who had intercourse orgasms with their partners had prolactin levels *four times* higher than those who orgasmed from masturbation. This extraordinary result was found for both men and women.

Compared to masturbation, the intercourse orgasm yields higher prolactin levels and therefore a feeling of great satiety. The fire is extinguished. You feel spent, in the best possible way. According to Brody, the more days each month you have penile-vaginal intercourse, and not any other kind of sex, the greater your overall well-being. However, the same study has yet to be done on anal and oral sex and masturbation with a partner, as well as intercourse without orgasm. (Obviously, your emotional satisfaction depends on other factors as well, including your connection to your partner.)

That orgasms from penile-vaginal intercourse are more satisfying than those from any other form of sex is a controversial assertion, but Brody has a theory. Deep penetration, combined with the heightened emotional and physiological aspects of intercourse, might more effectively stimulate the vagus nerve, which starts at the base of the brain and, running parallel to the spine, extends down to the chest, abdomen, and pelvic area. The vagus nerve supplies "electricity" to muscles and the sympathetic nervous system. It affects the heart, including the heartbeat, and is also involved in making you feel sated after eating. This brain-to-heart-to-genital live wire tunes the entire body. Some researchers believe the vagus nerve may stimulate the release of the soothing and bonding hormone oxytocin.

With or without the help of the vagus nerve, you get an oxytocin rush at orgasm, and the hormone might make you want to cuddle. Unfortunately, thanks to prolactin, your lover might not be up for it. Prolactin has another side effect, which seems to particularly affect men: it puts them in a sound, satisfied slumber.

Do men and women experience orgasm the same way?

In a classic study from the 1970s, psychologists at the University of Washington asked nearly fifty male and female students to write detailed descriptions of their orgasms, without mentioning genitalia. "A rush. Intense muscular spasms of the whole body," one person wrote. Another gushed, "An enormous buildup of tension, anxiety, strain followed by a period of total oblivion to sensation, then a tremendous expulsion of the buildup with a feeling of wonderfulness and relief." Reading this racy prose was a panel of seventy male and female gynecologists, obstetricians, and medical students. Their task was to determine if the author of each passage

Have Sex Before Stressful Events

If you have a stressful event coming up, plan to have sex beforehand. In a study on sex and stress, medical psychologist Stuart Brody asked twenty-two men and twenty-four women to give speeches in front of an unsupportive panel, followed by a test of verbal arithmetic. Each person's blood pressure was measured before and immediately after the speech. It turned out that men and women who had an orgasm from penile-vaginal intercourse, and this form of intercourse only, in the week or so prior to the speech had significantly lower blood pressure and reported less stress than people who had any other form of sex, including a varied schedule of intercourse and masturbation and/or sex with a partner that didn't involve vaginal intercourse.

You might wonder why masturbation, oral sex, and anal sex would detract from the alleged benefits of vaginal intercourse. There are no clear answers, and further research is required to replicate the result. Brody suggests it may have to do with how our bodies are "tuned" for up to a week after intercourse (affecting the vagus nerve, involved in soothing the heart), and the beneficial effects are disrupted if you stimulate other pathways.

was male or female—and they could not do it. Judging from the descriptions, the subjective experience of the orgasm sounded exactly the same for both genders.

Of course, there are a few obvious physical differences behind men's and women's moments of rapture. The vagina and penis have different physiologies. Women's orgasms comprise three to fifteen rhythmic contractions, lasting about fifteen seconds, although some orgasms go on for as long as two minutes. Women may also be multiorgasmic. The average male orgasm (ejaculation) is about ten to fifteen contractions and peters out after about seventeen seconds or so. Some compare the sensation of male orgasm to sneezing and the female orgasm to shivering. Both sexes experience an increase in heart rate, breathing, pupil diameter, and blood pressure.

Then there's the cognitive orgasm. For women, orgasm hits the pleasure spot known as the nucleus accumbens, and in men, it hits a connected structure known as the ventral tegmental area (VTA), which produces the feel-good hormone dopamine (see page 283 for more on the VTA). Both sexes also get a surge of oxytocin at the moment of orgasm, which may have a bonding and soothing effect. Right after orgasm, you get a prolactin rush, which contributes to a feeling of satiety.

Orgasms are anything but mindless. When neuroscientists look at the brains of people orgasming, they see activity in the region called the insula, known for processing emotional experiences and desires and translating them into conscious feelings. The insula might also have memory related to past sexual experience and the anticipation of reward. (Eating chocolate turns it on, too.) The parietal lobe is switched on as well; this region controls spatial navigation and is related to abstract concepts such as self-representation and self-expansion. Also activated during orgasm is the prefrontal cortex, an area associated with higher-level cognitive functions involving judgment, expectation, goals, and social control. Your ardent thrusts and spasms are made possible by the cerebellum, aka the "reptilian brain," the region that controls movement. Meanwhile, the left amygdala, associated with anxiety, shuts down, and so does the temporal lobe, involved in speech and hearing.

This is not to say there aren't possible differences between the male and female orgasmic experience. Men show more activity in the occipital cortex, the visual processing center of the brain. (If a couple uses mirrors and porn during sex, they're more likely for the man's pleasure.) Women but not men show activity in the hypothalamus and the medial amygdala, centers associated with emotion, drive, and the release of sex hormones, including oxytocin. (However, men do show activity in these regions during the arousal stage, so the absence of activity at orgasm could be due to technical difficulties in the PET scan. It might not capture everything that happens at orgasm, and requires subjects to keep their heads still.)

Unless you can lurk inside the brains of others as they climax, it's impossible to know *exactly* how their orgasms feel. But if you could feel the orgasm of someone of the opposite sex, you'd probably think the basic sensations of euphoria and release are familiar. Perhaps you can think of the orgasm as joy—there's no such thing as male joy and female joy. In fact, there's probably more variation within each gender than between them. It's how you get to the climax and how you feel afterward where a woman and a man are more likely to differ. At the peak it all looks basically the same.

Keep Your Feet Warm

In one orgasm study at the University of Groningen in the Netherlands, men and women were instructed to take turns lying down with their heads in a scanner, while their partners brought them to climax. To create the right mood, the doors were shut and the lights dimmed. The only problem was that the room was cold. Only half the couples were able to make it to climax. But when the researchers gave people socks to wear, the orgasm success rate shot up to 80 percent. The moral: if you really want to reach your climax, don't get cold feet. (Of course, you don't actually need to wear socks; just keep your peripherals toasty.)

Why do people with satisfying sex lives still masturbate?

Waxing philosophical, you might wonder why people who have plenty of sex with their partners would bother to masturbate. This especially applies to men, who are more frequent "self lovers." (According to a large-scale survey of American adults, 55 percent of men say they do it on a regular basis, at least several times a month, compared to only 38 percent of women.) Why, when sex is supposed to be about procreation, do so many of us masturbate so often?

Some answers are obvious. People masturbate to relieve boredom, as a form of safe sex, to induce sleep, for variation and convenience, to overcome sexual dysfunction, to reduce stress, and for the relief of burning urges. In the context of a relationship, couples consider masturbation more like a vitamin than a meal: in moderation, it doesn't hurt, and it might even help.

Those have been the existing theories on why we masturbate, and not many scientists have bothered with new ones since figuring out it's not the cause of blindness and hairy palms. Two exceptions are the British biologists Mark Bellis and Robin Baker. Contemplating the masturbatory habits of other animals, Bellis and Baker proposed an evolutionary explanation of why men masturbate so much—it gives them an edge in getting women pregnant. Yes, it's true that guys who masturbate regularly have less sperm in their ejaculate, which you might think would decrease the chances of conception. But fewer sperm can actually be a good thing. It turns out that the stuff has a "shelf life," and masturbation flushes away old stock. A man who has recently masturbated before making love would ejaculate younger, zippier, more competitive sperm that are more likely to get a woman pregnant. They'd live longer in her vagina, which means they'd have more time to swim around in search of a ripening egg. They'd also be more fiercely competitive against rivals' sperm in the event that the woman has more than one lover.

To test this theory, Baker and Bellis asked their female subjects to collect "flowback"—the pearly globules of sperm and vaginal fluids that flow out of the vagina after sex. They compared the sperm count in the flowback from women whose partners had recently masturbated before intercourse to that of women whose partners had not recently masturbated. Strikingly, they found that women retained just as many sperm after having sex with a man who had recently masturbated, even though there were fewer sperm in his ejaculate. Fresher sperm stayed in, and older sperm flushed out. (The actual sperm count depends on many factors, including the man's age, health, genetics, and how many days or hours before intercourse he masturbated. Men refill their tanks quickly, at an average rate of 2.41 million sperm an hour.)

For women, masturbation is also a way to "self-maintain." It stimulates blood flow to the genitals, maintains vaginal elasticity, and increases secretions. In this context, there's some truth to

Woody Allen's other meditation on masturbation: "I'm such a good lover because I practice a lot on my own."

Why aren't you sexually attracted to people who grew up with you?

Chances are if you grew up with a guy, you think the very idea of kissing him on the mouth is repulsive. Even if you find him attractive, it's probably not in a sexual way. According to anthropologist Debra Lieberman at the University of California at Santa Barbara, that's because you have a built-in kinship mechanism. When you were a child, this reflex kicked in if you saw your mother caring for a new baby, or if you lived with other kids in the same household or commune. The longer you lived in close proximity to other children, the more significant the kinship effect. Many other animals have a kinship mechanism, too, and it serves an important evolutionary purpose. If parents are closely related, there's a higher chance that offspring will be born with genetic disabilities. Inbreeding is a genetic no-no.

You can't override your incest-avoidance mechanism by knowing that the person reared with you is a stepbrother, adopted brother, friend of the family, or otherwise not related genetically. It's an indelible neural imprint. Evidence for this has been found in studies of Israeli kibbutzim, where children are raised communally in small groups. In a survey of nearly twenty-eight hundred marriages, only fourteen were between kibbutzniks who grew up together in the same peer group, and all but five of those couples hadn't lived with each other in the first six years of their lives. A similar phenomenon happened in China, when baby girls were adopted into the families of the boys they'd later marry. Because these boys and girls grew up together like brother and sister, they developed a sexual aversion to each other. Compared to Chinese women in other arranged marriages, they had twice as many extramarital affairs, a significantly higher rate of divorce, and a 30 percent lower fertility rate.

Researchers don't yet know which specific environmental cues, such as touching, kissing, eating, or sleeping together, might trigger our kinship detection system, or the exact genetic basis and neural mechanism. But since in most situations children growing up together are in fact genetically related, from an evolutionary standpoint the loss of a potential mate is less costly than the risk of inbreeding.

The bottom line is that you've evolved to keep your sex life outside the family circle. Which is why telling a guy you love him like a brother is code for saying there's absolutely *no way* you'd ever sleep with him.

PART III
BRAINS

CHAPTER 8 *The Dating Mind-set*

Are you the new person drawn toward me?
To begin with, take warning, I am surely far different from
what you suppose;
Do you suppose you will find in me your ideal?
Do you think it so easy to have me become your lover?
Do you think the friendship of me would be unalloy'd
satisfaction?
Do you think I am trusty and faithful?
Do you see no further than this façade, this smooth and
tolerant manner of me?
Do you suppose yourself advancing on real ground toward
a real heroic man?
Have you no thought, O dreamer, that it may be all
maya, illusion?

—Walt Whitman, from "Are You the
New Person Drawn Toward Me?"

What do women and men value in a partner?

Seen through an evolutionary lens at least ten thousand years thick, it's perfectly natural that, like the Marilyn Monroe character in *Gentlemen Prefer Blondes*, women have a thing for men with money and status and men have a thing for women with good looks. The story has been told over and over again: Women evolved a preference for partners who are good protectors and providers. Men evolved to invest their resources in women who have qualities related to fertility. The startling thing is, even in the twenty-first century, these mating biases not only hold true but also are basically intact *everywhere around the world*. Times have changed, but we haven't.

A landmark study by evolutionary psychologist David Buss arrived at this conclusion after tallying the mate preferences of more than ten thousand people in 37 cultures around the globe. From America to Zambia, women value "high earning potential" and "good financial prospects" in a partner, much more so than men do. This is especially true in countries such as Indonesia, Nigeria, Japan, and Zambia, where women have few financial prospects. But it's also true throughout the Americas and Asia, and is statistically significant even in the most egalitarian countries of Western Europe (with the exception of Spain, where women still care about men's resources but not significantly). On a related note, women value ambition and industriousness in men more than men do in women (the significant exception are the South African Zulu, where women take on most of the work of building a house and fetching water while men must travel long distances to find work). A guy's earning potential not only might provide women with status and protection but also could indicate that he has "good genes" that could be inherited by their children.

Meanwhile, unsurprisingly, men throughout the world care more than women do about their mate's age and looks. In every one of the thirty-seven cultures, men prefer their wives to be younger. On average, men prefer to get married at twenty-seven and a half to a twenty-five-year-old woman. Women also prefer their husbands a little older and presumably more established; in fact, a three-and-a-half-year age gap is considered ideal. If the man divorces and remarries, the age gap grows to five years at second marriage and eight years at third marriage, probably due more to men's desires than to women's. For women, an older man is likely to have more resources. For men, a younger woman is probably more fertile.

Men's emphasis on youth and fertility goes hand in hand with their general preference for good looks: lustrous hair, clear skin, symmetrical features, a low waist-to-hip ratio, and so on. In thirty-four of the thirty-seven cultures surveyed, men cared much more for a physically attractive spouse than did women. Only in India, Poland, and Sweden did men reportedly not value their wife's looks all that much more than women cared about their husband's.

(Why these three countries? In India, families arrange marriages. Poland and Sweden are harder to explain; perhaps every woman is a beauty.)

How much a woman desires a good-looking partner may depend in part on her own looks. A study of nearly 150 Polish women found that women with a high WHR (thicker waist) valued physical attractiveness in a partner more than did women with a low WHR (curvaceous body type). (Curvy, feminine-looking women instead placed more emphasis on a prospective partner's resources and earning potential.) The researchers, biological anthropologists Boguslaw Pawlowski and Grażyna Jasieńska, suggest that when a woman isn't considered good-looking, she may subconsciously value genes related to attractiveness to increase her children's chances of reproductive success. That said, a study on marital happiness by psychologists at UCLA found that when the husband is more attractive than the wife, the marriage is more stressed. The best predictor of a good marriage, they concluded, is when the wife is much hotter than her husband. Presumably, the husband works harder in the relationship and is less likely to cheat, and the wife feels more secure.

Even so, the most beautiful women want it all in a man, according to David Buss and Todd Shackelford, in a study on American married couples. The evolutionary psychologists recruited a rotating team of male and female interviewers who paired up and evaluated more than two hundred married participants in the Midwest. Each subject was judged for physical attractiveness and assessed in three separate sessions for the factors they valued and insisted on in choosing a mate. The prettiest women had the highest standards—they wanted and expected their partners to be masculine, fit, physically attractive, loving, educated, a few years older than themselves, and desirous of home and children, with a high income potential. Surprising to the researchers, there was only one quality beautiful women did not insist on more than plainer women did: intelligence. (The hottest men, meanwhile, didn't have higher standards than homelier men. A wealthy guy, on the other hand, may be choosier.)

Buss and Shackelford concluded that while exceptionally attractive women might get it all in one man, the majority can't both attract and retain that ideal guy, so they settle for the best combination of traits in one partner. Maybe they forfeit masculinity for paternal proclivity, or trade resources for loyalty, depending upon individual choice and circumstance. A minority of women adopt a "dual mating" strategy, which means benefiting from a long-term partner and having trysts with men who have desirable physical or behavioral qualities that their primary mate lacks.

As obsessed as men are with youth and looks, they do look for other criteria in a wife. In a study by evolutionary psychologist Norman Li and his colleagues at Arizona State and Northwestern universities, adult subjects (mostly in their thirties) were given a low number of "mate dollars," which they had to "spend" on various characteristics such as income, friendliness, or an interesting personality. Forced to choose the qualities most important to them in a partner, men valued physical attractiveness the most, closely followed by intelligence. Intelligence, according to the researchers, is something of an umbrella term, covering qualities such as parenting capability, social savvy, and the ability to run a household or have an interesting conversation. (Remarkably, women in this study considered intelligence the most essential quality in a spouse, followed closely by yearly income. The finding that all women value intelligence so much could explain why it's the only quality in a mate that gorgeous women don't insist on any more than plainer-looking women do.) When the researchers introduced kindness as a characteristic in a follow-up study with undergraduates as subjects, men and women alike chose it as the second most important quality they value in a partner (after physical attractiveness and social status, respectively).

The bottom line in these studies is that men value beauty and youth in a partner more than women do, and women parlay their looks into getting the mates who can offer them the most. There's some hope for humankind in that we also claim to value the qualities of kindness and intelligence in our long-term relationships. Too bad that doesn't prevent one out of every two marriages from combusting.

What secret biases do data from online dating sites reveal?

Researchers love online dating services because they get a data gold mine on who attracts whom, and why. It's real-life data, the stories of real people who have made real decisions. It's based on action, not theory. Imagine, then, how delighted the team of MIT and University of Chicago economists and business school professors must have been to have access to the preferences of more than twenty-three thousand singles in Boston and San Francisco who used a major online dating site. Spelunking through the data, they unearthed the unspoken biases that men and women don't want to admit, not even to themselves.

Unsurprisingly, one of the big prejudices that surfaced concerns attractiveness. Yes, you know it already: men care significantly more about looks than women do. But now we know how much they care, in quantifiable ways. For instance, you're twice as likely to get a man's attention if you post a photo of yourself online. Between your photo and self-description, prospective dates get a sense of your face, hair, and body, which they evidently care about quite a lot. Judging by the number of first-contact e-mails, men love long straight hair. They prefer your hair color most if you're blond, and least if you've gone gray or salt-and-pepper. Regardless of their own height, men were biased against women over five feet nine.

Men also prefer women who are light for their height. If you had a body mass index (BMI) of 16–18, the weight-to-height ratio of an anorexic (approximately 100–112 pounds at five feet six), you'd have 90 percent more first-contact e-mails than a woman with a BMI of 24, which is at the high end of the normal range. Interestingly, the average weight of a woman using the dating service was six pounds lower than the national average for women under thirty, eighteen pounds lower for women between thirty and thirty-nine, and twenty pounds lower among women forty to forty-nine. A male friend who's a seasoned online dater told me that of course all men know that women lie about their weight in their profiles. "There's the fudge factor," he said, giving me a bitter smile. "If a guy says he prefers a BMI slightly lower than average, he knows he'll actually get a woman of average weight. If a woman says she's average weight, it means she's a little pudgy." Indeed, a subsequent study at Cornell University on deception in online dating found that women do underreport their weight; 64 percent of women shed five pounds or more in their online profiles.

Meanwhile, women homed in on the part of men's profiles that specify their income. Guys who said they earned an annual income of $250,000 or more received 34 to 151 percent more first-contact e-mails from women than their peers who made an average salary of $62,500. (Salary inflation is apparently as common as weight deflation.) How much more interest a big breadwinner received than the average salaried worker depended on his physical details such as looks and height.

The researchers used a simulation to predict outcomes for various scenarios. Based on dating patterns, if a guy's physical appearance is rated among the least attractive 10 percent of men on the site, he'd need to make an additional $186,000 each year to be considered equally desirable to a great-looking stud with an average income of $62,500. If a man is only five feet six inches tall, he'd need to make an additional $175,000 to be as desirable as a man who is five-ten and average in every other way. Of course, guys who present themselves as tall, rich, and good-looking get the most attention. (FYI, men also lie about their height; 53 percent stretch the truth a half inch or more in their profiles.)

Women are much pickier about race than are men, preferring their own skin color to any other. Even when ladies claim they have no ethnic preference, they do. Americans who grew up south of the Mason-Dixon line were more biased in favor of their own race, according to a speed-dating study at Columbia University led by economist Raymond Fisman and psychologist Sheena Iyengar. In their set of four hundred speed-daters living in New York City, black women were the choosiest, strongly preferring black men over guys of any other race. To the prospect of a second date, black women said yes approximately 65 percent less often to Asian men, 45 percent less often to white men, and 30 percent less often to Hispanic men. White women preferred white men, and said yes 65 percent less often to Asian men and 30 percent less often to black and Hispanic men. Hispanic women preferred Hispanic men, and said yes 50 percent less often to Asian men and 20 percent less often to black and white men. Asian women were the least racially discriminating, saying yes to Asian men and white men only slightly more often than to black and Hispanic men.

Once again, men's income can overcome women's biases, according to the University of Chicago and MIT professors who analyzed the online dating data. For white women, a black man would need to earn $154,000 more each year to be considered equally desirable as a white guy with an average income of

$62,500, Hispanic men an additional $77,000, and Asian men an additional $247,000. For black women, a white man would need to earn an additional $220,000 to be equally desirable as a black man, and a Hispanic man would need to earn an additional $184,000. For Hispanic women, a black man would need to earn an additional $30,000 to be equally desirable as a Hispanic man, and a white man would need to earn an additional $59,000. Asian women were the least racially discriminating.

(The studies don't provide insights on why non-Asian women preferred Asian men the least, but women are happy to speak off the record. In my informal survey, I received answers that ranged from a perception of Asian men not having the same interests and values, a belief that Asian men have close-knit communities and don't marry outside their race, alleged sexism in some Asian cultures, and the perception that Asian guys are generally shorter. Women are sheepish about stereotyping, and say of course they'd make an exception for the right man.)

Although a big paycheck can compensate for a man's perceived shortcomings, women also care about *how* a guy makes his money. Men who are lawyers, firefighters, military personnel, and in health-related professions (doctors and surgeons, probably) attract more women. Guys with manufacturing jobs received fewer e-mails from women. Meanwhile, men are basically indifferent about women's income or profession, as long as a woman's success doesn't intimidate them. In fact, men contact female students slightly more often than women with jobs. The Columbia University study showed that men may like smart women, but not if the women appear to exceed them in intelligence or ambition. Given these data points, it's no wonder we hear about female surgeons and lawyers saying they are nurses or secretaries just to get a date, and how, when they bring the man home, they hide their designer shopping bags and other signs of affluence.

The researchers also found an assortment of other biases that are unfortunate but unsurprising. Men turn up their noses at older women, and older women turn up their noses at younger men. Women prefer men with equivalent education levels, and men who have only high school diplomas avoid women who have more education. Women with children are less desirable to men, whether or not those men have kids, and childless women prefer men who don't already have kids. Women who are divorced prefer fellow divorcés, whereas divorced men prefer women who have never been married.

Because there are far more people online than at any one bar or club, and because online dating is so safe and so abstract, it's easy to be superselective. But the number-crunching researchers haven't yet drilled deep enough to know how personality affects our chances, how the most finicky of us do on dates, how many end up in marriages, and how many of those marriages are happy. Back to the trenches.

Don't Judge a Book by Its Cover

If you're attracted to someone, why not make an attempt to know that person rather than assume he doesn't have anything in common with you? Professors Raymond Fisman and Sheena Iyengar at Columbia University, along with their colleagues at Harvard and Stanford, encountered racial discrimination in their experimental speed-dating study. To their surprise, even within an educated and relatively progressive community of graduate students at Columbia University in Manhattan, there were stark racial biases among female singles. Women said the reason they prefer same-race men is they feel they have more in common with them. Totally unfair, you might say—how much can a gal tell about a guy's tastes in four minutes? The answer is a lot—provided he brings props. Remarkably, the researchers found that women's same-race preferences weakened when they told men to bring a favorite magazine or work of classical literature with them on the speed date. It appears that the reading material became a topic of conversation and helped women dismiss the idea that they didn't have anything in common with men of other races. It probably brought to mind the universal rule: you can't judge a book by its cover.

Why might single men spend more and single women volunteer more?

Men and women have different ways of showing off their goods, according to evolutionary psychologist Geoffrey Miller and doctoral researcher Vladas Griskevicius. Men display their status and resources through conspicuous consumption—buying luxurious cars, homes, and gifts. Meanwhile, women flaunt their tendencies through conspicuous benevolence—self-sacrifice or charity. After all, the number two quality men seek in wives, after physical attraction, is kindness and one of the top qualities women seek in husbands is social status/resources.

The psychologists decided to test this theory by splitting male and female participants into two groups. One group was "romantically primed" with pictures of attractive opposite-sex singles. They were told to pick a hottie from the selection and write about what would happen on a date with that person. The other group, a control for comparison purposes, looked at pictures of buildings and were instructed to write about the weather. Following that, both groups were told they had $5,000 in their bank accounts and asked to indicate how much of it they'd spend on luxuries such as a watch, car, or European vacation. They were also told they had sixty hours of free time each month to help at a shelter, build housing, or work at a hospital, and to indicate where they'd volunteer and how many hours they'd donate.

The difference between the "romantically motivated" group and the control group was striking. The subjects in the control group reported little interest in volunteering or spending money. Not so of the people in the romantically primed group: the men said they'd spend more money on stuff (and time, but only if it meant doing something heroic, such as saving lives), and the women said they'd volunteer more often.

According to the researchers, the display of wealth or benevolence is useful in attracting mates only when others can observe it. It's part of the mating mind-set. Because the sexes generally value different qualities in a long-term relationship, women benefit most from advertising themselves as helpful and self-sacrificing, whereas men display their wealth and heroism.

Cultural expectations, including religion and social standing, no doubt have a lot to do with why men spend and women give. But evolutionary theory provides further insight, given that women seek husbands who have resources or status, and men, although less choosy, prefer wives who are not only attractive but also good, giving mothers. All this casts new light on everything from charity balls and firefighters' parades to rosters of benefactors. Very few people seem to donate time or money anonymously. If they do, they're often put into a separate, asexual category: they're called saints.

Go on a Volunteer Date

As an alternative to a candlelight dinner, why not try the volunteer date? Find an appropriate activity, such as planting a community garden, serving lunch at a shelter, or taking a kid on an expedition to a museum. You get to show off your good heart, gain insight on how your date acts in a real-life setting, and do some good. (You can try to meet someone while volunteering, but keep in mind that there are far more do-gooding women than men.) Volunteer work also gets the feel-good endorphins flowing through your bodies, puts you in a fantastic mood, and creates an intimacy that can carry over in the evening when you both need a back massage. The Web site www.volunteermatch.com lists volunteer activities by zip code.

Why do men give women fancy dinners and vacations instead of useful gifts?

To woo a female dance fly into having sex, the male fly has to offer a nuptial gift, which is, ideally, a whole tiny insect or at least a silk-wrapped fragment of one. The bigger the gift, the longer the copulation. It's not exactly like this for men and women, but there are parallels. Both women and female dance flies invest more resources than males in offspring, so they care about picking the best possible mate. Gifts from males are signs of their resources and intentions. If the male gives costly gifts, he's sending a signal that he's honorable and, if he's human, that he has a long-term interest. If he gives cheap (or thoughtless) gifts, he's not that into you, or he fails to impress.

From a human male's perspective, the problem with gift giving is that women might be using him just for the goods. How does a guy send the right message and get what he wants without being strung along? This dilemma interested British mathematician Peter Sozou, who heard about a man who had been paying his girlfriend's rent only to find out that she had been cheating on him with other men. Sozou and his colleague Robert Seymour at University College London took a game theory model known as a Nash equilibrium (named after its creator, mathematician John Nash, profiled in the movie *A Beautiful Mind*) and applied it to optimal gift-giving strategies.

The Nash equilibrium is a mathematical model that refers to a situation in a game when players pursue the best possible strategy for themselves, maximizing their own self-interest, while taking into account the decisions and strategies of the other players. When the Nash equilibrium is applied to courtship, where the evolutionary and social pressures on men and women are a given, the best strategy for men is to show off their spending power and buy women costly gifts that suggest good intentions. But there's a limit: the swag should be worthless in the sense that it should lack value outside the context of courtship. Logically, it serves men best to wine and dine you and take you to Hawaii rather than give you cars and computers or pay your bills. If you're not genuinely interested in the guy, chances are you won't accept an infinite number of dinners, plays, and trips with him. Uninterested women have little incentive to accept worthless (or experiential) gifts, and men don't have to invest as much in wooing unwilling women. This strategy separates legitimately interested women from gold diggers.

Even without fancy game theory to explain its virtues, conspicuous consumption has long been the diet of romance. Flowers, chocolates, drinks, dinners, and trips are tried-and-true courtship rituals. From an evolutionary perspective, women prefer men who spend extravagantly, whether frivolously or practically, because they are displaying resources and showing signs of good intentions. The more invested, the stronger the signal that he won't desert her and their potential children. Whimsical gifts may in fact be better because they reveal a general willingness to invest resources in a woman, unburdened by practicality. (Jewelry apparently falls somewhere in the middle, having less practical value than houses and cars.)

Like most mathematical models, however, reality intervenes. There are some women who feel uncomfortable receiving gifts or find them manipulative or antifeminist. There are others who'll string along a beau just to go on trips and outings. If that's the case, perhaps they'll meet up with their male counterparts—well-endowed playboys who give extravagant gifts of trips and dinners to multiple women, without any good intentions.

Why does creativity get men laid?

Creativity—whether displayed in art, music, language, humor, or novel ideas—has evolved at least in part from the male drive to, well, get laid. According to Geoffrey Miller in his book *The Mating Mind*, choosy women are attracted to men with vivid brains much in the same way that peahens are attracted to peacocks with vivid tails. Over the generations, peacock tails have become bigger and brighter than necessary for mere survival, and so in their way have human brains—and both owe their existence to lusting males vying for the sexual attention of discerning females. As Jean-Paul Sartre admitted, "If I became a philosopher . . . it's all been to seduce women basically."

Unconsciously, we desire men with good genes ("sexy genes") that may be passed on to our children. It follows that we look to men's creative expression as an indication of traits that may be at least partly genetic: intelligence, willingness to grow, problem solving, flexibility, risk taking, perseverance, cultural fluency, and strengths in the verbal, numerical, perceptual, spatial, memory, social, or emotional realms. If a man shows off his brainpower by making you his muse—by sweet-talking, writing you a romantic note, telling jokes well, singing a song, or spinning a story—you might be smitten. According to Miller, it's no coincidence that the majority of men's creative work is produced between ages twenty and thirty-five, their most fertile years.

Creative men, as a group, may also be more prolific lovers. Surveying more than four hundred British poets, visual artists, and other creative types, anthropologist Daniel Nettle found that artistic men and women who report having "unusual experiences" and "being impulsive nonconformists" have a higher number of lifelong sex partners and perhaps a stronger sex drive than people who aren't creative. They might have an especially easy time seducing a woman into bed when she's in the fertile phase of her menstrual cycle. In one study at UCLA, women were asked to read two vignettes, one about a creative but poor man and the other about a noncreative but rich man, then choose which man they'd prefer. For a short-term relationship, women who were ovulating at the time of the study significantly preferred the destitute creative guy over the dull rich one.

Creativity is clearly sexy, and it gets men laid. But it's tough to gauge how much women really value creativity in a long-term relationship, because it overlaps with intelligence, social status, humor, and other sexy qualities that women care about a lot. In one study of mate preferences led by evolutionary psychologist David Buss, creativity ranked somewhere in the middle. Undergraduate women considered it less important than kindness, handsomeness, and healthiness, but more important than good heredity and (ironically) being a college graduate. Then again, it's no shock that

American college students value creativity given that they live in a culture that gives high status, at least in theory, to artists, entertainers, and entrepreneurs.

However, when nearly eighty middle-class, middle-aged Americans were asked how much they value creativity in a partner, they had totally different priorities. Evolutionary psychologist Norman Li and his colleagues at Arizona State and Northwestern universities told subjects they could "customize" their ideal partners. The challenge was that they had a limited "mate budget" and were forced to divvy up their "mate dollars" according to the qualities that mattered most to them. When the mate budget was very low, women and men ranked creativity at the bottom, but when each player got more mate dollars to spend and requirements such as personality, romance, and a basic income level were satisfied, women's budget for creativity increased by 8 percent. (Men spent more mate dollars on a partner's creativity, especially when their budgets were tight, in part because they didn't budget as much as women did on income or work ethic.) Women value creativity much more when their "budget" is big and other mate criteria are fulfilled.

Could it be that creativity is a luxury and so you're more attracted to creative types when you're younger, when you come from an affluent background, when all your practical needs are met, or when you just want a passionate fling? Maybe; the researchers seem to think so. Yet so much is individual and depends on circumstance and happenstance. As the great artist and lover Pablo Picasso observed, "One never knows what one is going to do. One starts painting and then it becomes something quite different." The same can be said of dating, which is in itself an art.

Why aren't there more male muses?

Just when the poet Edna St. Vincent Millay's marriage was ostensibly its most stable, she fell passionately in love with George Dillon, a Pulitzer Prize–winning poet fourteen years her junior. She wrote him poems, praising "this love, this longing, this oblivious thing." Her intensity both attracted and repulsed the young man; they'd rendezvous, he'd retreat, and she'd advance again. After one of their many heated breakups, she went on an inspired tear, producing *Fatal Interview*, a collection of poems the critics deemed her very best. She wrote it to woo him, to heal her wounds, to keep him in her clutches.

George Dillon should have felt very special. Evolutionary psychologists have puzzled over why, with the sexes being equally creative, men seem to use their creative talents more often than women to lure potential lovers. What would Dalí, Picasso, Dante, and Nietzsche have accomplished without their many mistress-muses? Where are all the mister-muses?

The answer, from an evolutionary perspective, is that women have less need to display "good genes" to men than men do to women. Most men aren't as choosy as women are when selecting mates, and guys respond more strongly to sexual and aesthetic cues than to creative ones. While guys do value creativity, humor, and intelligence in a woman, it's usually in the context of a lasting relationship. Therefore, women are more likely to save their creative displays for guys who have long-term potential.

This theory, anyway, was tested in a study on the effects of romantic motives on creativity by Arizona State University psychologists Douglas Kenrick and Robert Cialdini and researcher Vladas Griskevicius. They recruited more than six hundred men and women and asked them to do various creative writing tasks and take a creativity test. Some of the subjects were primed with various romantic scenarios—involving either short-term relationships, such as a fling on the last day of an island vacation, or long-term relationships, such as a lasting college romance—while a control group was not primed with thoughts of love. As predicted, the difference between men and women was dramatic. Men displayed increased creativity when primed by any sort of love affair, whether a brief tryst or a lifelong love. The women, however, tested higher on a creativity test only when primed with the idea of a long-term relationship with a sensitive and devoted man. The idea of a hot fling on the last day of vacation might have tickled their imagination but didn't move them to be more creative. Judging by these results, women prefer their muses to be more like permanent installations than passing peep shows.

But sometimes the muse flits off, as George Dillon did to Edna St. Vincent Millay. So Millay poured that passion into her last tremendous epic work. She didn't have a muse or a mister in the end, but she had a masterpiece.

> *Well, I have lost you; and I lost you fairly*
> *In my own way, and with my full consent.*
> *If I had loved you less or played you slyly*
> *I might have held you a summer more,*
> *But at the cost of words I value highly . . .*
> —Edna St. Vincent Millay,
> from "Fatal Interview"

Why is humor a turn-on?

If there's anything that cuts through discomfort, it's humor. Case in point: You're on a first date and the guy tells you a story about his family, doing impressions of his crazy grandfather. It cracks you up. Playacting, you claim to covet the vintage watch he's wearing. He laughs and assumes the role of an oily salesman trying to sell it to you for more than it's worth. You pretend to be a naive buyer who's getting swindled. You both laugh and eye each other mischievously.

Humor works on many levels. It provides psychological distance from practicalities, allows us to take ourselves a little less seriously, and reduces the tension of a first encounter. It's also a powerful bonding agent. Whether it's your witticism or his, you're sharing something about yourself and sending a signal of appreciation and solidarity.

To test humor's impact on attraction, psychologists Barbara Fraley and Arthur Aron at the State University of New York at Stony Brook recruited nearly a hundred people who didn't know one another, randomly paired them up, and asked them to participate in a series of joint assignments. Half the pairs performed tasks that encouraged them to be funny, such as performing stunts while blindfolded and speech-inhibited, acting out TV commercials in a language they made up on the spot, playing charades, and so on. The other half of the group was given assignments that required them to work together but didn't elicit humor. When the participants were later asked their opinions of their partners, those who laughed together felt significantly closer and more physically attracted to their partners than those who did not.

Although both genders rate a good sense of humor as one of the most desirable qualities in a mate, women more often say they want a partner who can make them laugh, while men seek a gal who "gets their sense of humor." Women also rank humor higher than men do on their lists of desired traits in a spouse. Evolutionary psychologist Geoffrey Miller suggests that this makes sense from an evolutionary perspective. Guys show off their "sexy genes"—signs of intelligence and personality, which may be partly heritable—while women, the choosier sex, are amused and impressed, or not. Psychologists have also found that if a man and woman are laughing together, her laughter is a better indicator than his of romantic interest.

Could it be that men and women process humor differently and that women are by nature a more responsive audience? Stanford University neuroscientists tackled the question by using fMRI to track the activity in men's and women's brains as they went about the task of viewing and rating cartoons. Both sexes, it turned out, were similar in their selectivity and response time to humor, but women showed more activity in the left prefrontal cortex, the region responsible for executive language processing and judgment, and the nucleus accumbens, the brain's "reward center." This suggests that women could process jokes in a more complex way than guys do (which may include judging them more) and may even appreciate humor more or find it more stimulating. It's also possible that women evolved to find humor more rewarding because it motivates us to seek out men with good genes.

While men may use humor more often than women do to attract dates, both sexes value it throughout a relationship. After the courtship stage, the humor balance may level, with both sexes making each other laugh. (Think of Myrna Loy and William Powell, the hilarious husband-and-wife duo in the classic Thin Man films, or Katharine Hepburn and Spencer Tracy.) A witty repartee can keep a couple together long after the blushes and giggles fade. Humor is, as Fraley and Aron describe it, a process of self-expansion. It's an escape hatch into a sidesplitting alternative reality, if only for a moment.

Share a Sense of Humor

Find a partner who shares your sense of humor, joke often, create running jokes, and frequently bring up your funny moments together. According to a study by psychologist Doris Bazzini at Appalachian State University, couples who create inside jokes and reminisce about them together are more satisfied with their relationships than couples who don't. Not only are inside jokes a way to bond, they also release endorphins that make you feel good.

This lesson became real to me when, many years ago, I volunteered to work with Alzheimer's patients and their spouses. It was heartbreaking for these spouses to see their loved ones fade away. However, a couple's mutual sense of humor often remained intact even after the Alzheimer's patient could no longer remember his or her partner's or children's names. Humor was a small mercy in this tragic degenerative disease, and the last thread that attached spouses to their loved ones. (I remember how one husband with Alzheimer's lovingly called his wife the zookeeper and himself the monkey or the tiger or the snake, depending on how he felt that day. She'd giggle and play along.) Studies show that people value humor highly in a long-term relationship, but this was living proof of how deeply ingrained humor is between a couple; it not only attracts us to our partners but also may keep us together after much else has unraveled.

Why are you more attracted to picky people (and they to you)?

For the 150 hopeful hearts who participated in four-minute speed-dating sessions at Northwestern University and MIT, success depended on how discriminating they were. Anyone who walked into the room with an open-minded romantic interest in many of the people there was likely to leave without a second date. What a single man or woman might have considered a warmhearted approach to the dating process was subconsciously perceived by others as desperate, flirtatious, or, if at all appealing, then only for friendship. The lesson is that no one wants to embrace you if you're all-embracing.

The study, led by psychologist Eli Finkel and researcher Paul Eastwick at Northwestern University, found that pickiness is good, and pickiness with a purpose is even better. Singles were more likely to get a second date if they homed in on *one* person only. The most successful daters make their target feel unique, even in the few moments of an encounter. It's like what people say of former president Bill Clinton: he'd zoom in on you and only you, making it seem as if you were the center of his universe while you were with him. The best politicians and lovers know that getting what you want comes from making people feel they're unique—one in six billion.

The desire to feel special applies equally to both sexes. Especially for women seeking a long-term relationship, promiscuousness has zero appeal. (Men, too, don't want to invest in wives who'll cheat.) If a guy is a womanizer, the type who flirts with every gal in the room indiscriminately, you're less likely to think he's Mr. Right. But if a man singles you out in a crowd and makes it clear that no one else interests him, you're likely to think he's pretty special, too. And even if he doesn't meet the standards you've set, you might make an exception, which often happens when speed-dating and in random encounters.

If you're picky, your interest boosts not only a man's self-esteem but his mate value as well. Mate value is a measure of how attractive you are to potential partners, and it reflects your personal history of dating and dumping and getting dumped. Think of mate value as a stock that goes up and down in the dating stock market. Your mate value goes up when you're perceived as selective and thoughtful, desired by many, date attractive people, and rarely dumped; conversely, your mate value goes down if you're less selective, less attractive, and often dumped. Have you ever noticed how once you start to fall in love, other opportunities seem to open up magically? Some of it is probably due to subtle changes in your behavior and hormones which make you more attractive. But it's also probably due to a spike in your mate value.

People seem to care a lot about mate value because it has everything to do with status and ego. We want to hear all about each other's pasts—how many people they've slept with, what the exes were like, the duration of the relationships, and so on. In fact, men care so much about mate value that, according to a study at the University of Texas at Austin, "figuring out my mate value" is the second reason why they have casual sex (after physical attraction). If a guy perceives you as having a high mate value, then he's more into you. After all, if you're attractive and discriminating, then naturally your mate value *must* be high, or so the logic goes. Too often we hear stories of a man losing interest in a woman, or belittling her, or not taking the relationship further after discovering she wasn't very selective in her dating history. Women are known to do the same to men.

The bottom line is that you want your suitor to be picky, and he wants you to be just as picky. It's like a twist on the Groucho Marx maxim: you don't want to be a member of any club that's too quick to accept you—and when you do, you only want there to be two members.

If I am not worth the wooing,
I am surely not worth the winning.
—Henry Wadsworth Longfellow

Make Your Date Feel Unique

My friend Zoë recently went on a date with a guy she met on a popular online dating site. This was the twelfth person she had met online, and by this point she was accustomed to the ritual exchange of dating histories. Zoë found these dialogues not only tedious but excruciatingly unpleasant; one man called her a "veteran" after she revealed how many unsuccessful dates she had been on, and she found it dispiriting to hear men rattle off accounts of their own dead-end encounters, sensing that she was about to become one of them. She felt run-of-the-mill, and these guys seemed like damaged goods. Her mate value seemed low and was getting lower as she was rejected by what she perceived as low-value guys—until this twelfth match. Her date, Dan, told her, "I could tell from the second minute I met you that this would be fun, unlike so many of the dates I'd been on." His remark changed everything. He singled Zoë out and gave her a high value by default, and she in turn saw that they had special chemistry. A self-fulfilling prophecy? Maybe. But they had a fantastic time, and not long after started to date each other exclusively.

Why do people seem hotter when others are into them?

Female zebra finches understand how crucial it is to pick the right mate, one who is resourceful and willing to help raise the little ones. They also know that finding him isn't easy. Sometimes a zebra finch decides that her best strategy is to follow the good judgment of other females. If she's forced to choose between two males, one solo and the other a recipient of female attention, she'll probably spring for the popular one, or at least another bird that resembles him or wears an armband of the same color.

Observing how female zebra finches copy one another's choice of mates, psychologist Ben Jones and his colleagues at the Face Research Lab at the University of Aberdeen tested women to see if they'd do the same. "Who is more attractive and by how much?" the researchers asked, showing pairs of men's faces side by side. The faces were prejudged as roughly equal in attractiveness, but the women usually indicated a slight preference for one or the other. In the second part of the experiment, the researchers showed the judges a slide show starring the same men's faces, but this time with a twist. A profile of a pretty woman was shown looking at one of the two men, and she wore either a smiling, happy expression or an unsmiling, neutral expression on her face.

The woman's expression made all the difference in how the

man was perceived. If the lady was smiling, female judges thought he was hotter, and gave him at least 15 percent higher ratings than in the first round. If she had a neutral, sullen, or bored look, women perceived the man as homelier and downgraded him by more than 10 percent on average. Choosing between two men, women consistently favored the guy who was the recipient of the smile. Women also thought men were much more handsome when paired with an apparent admirer than when they were shown by themselves. Handsomeness, it seems, is in the eye of *other* beholders.

The funny thing is that when the researchers asked male judges to rate the same male faces, the results were the very opposite. Men judged other guys as less attractive if a pretty woman was smiling at them, and boosted their rating of the guy if the woman wore a neutral expression on her face. It's male competitiveness, according to the researchers. Men are threatened by other men who soak up the attention of women. (This helps explain why some guy friends never approve of the men you date.)

You might also wonder how much the admirer's attractiveness counts. The answer is a lot, according to a subsequent study led by Anthony Little at the University of Stirling in England. Moreover, men appear to be influenced by other men, just as women are by other women. When Little and his colleague paired a picture of a woman with a picture of an attractive, masculine man, male judges gave her higher attractiveness ratings than when she was partnered with a feminine-looking man. The same was true for men: guys were given higher ratings by women when paired with good-looking, feminine women rather than masculine-looking women. Interestingly, the ratings were significantly higher only when the judges were rating the men and women as candidates for a long-term relationship, not a fling. It's reasonable to conclude that both men and women care more about social cues when judging people as candidates for serious commitment. For flings, we depend less on the opinions of others.

Why do we take our cues from other people? For the same reason the single female zebra finch takes her cues from other zebra finches, according to Jones and his colleagues. Finding the best mate takes time and energy, and it's not easy. By adopting a wisdom-of-crowds approach, you're more likely to focus your time and effort on a worthwhile candidate who has already been tacitly endorsed. If a guy attracts beautiful women and charms them, he's probably a decent catch. Likewise, if you've had a handsome or successful boyfriend, you might be more desirable to other men, especially in the context of a potential long-term relationship. Your "mate value"—desirability to others based on past dating history—is higher.

Researchers have yet to determine if, like female zebra finches, women would settle for guys who simply *look* like the guys that other females desire. (This might work for Johnny Depp look-alikes, but what about Woody Allen look-alikes?) We don't yet know if a woman who dates a handsome man is still considered more attractive if the guy has a reputation as a jerk, or if he cheats on her. Finally, you might wonder how long the improved perception lasts. Do you find a man attractive only for as long as other pretty women appear interested in him? Or do you dump him once he gets dumped?

Keep a Flock of Wingmen

We're attracted to people who have already received interest. That's true of women and men alike. It's probably why some clever women always seem to have sexually motivated guy friends around, or boast about their male admirers even if they have no real interest in them. If you have devotees, you're more attractive to other men and your "mate value" rises.

This also means you might get a lift by bringing a friend with you to a social setting such as a party or club. A "wingman" is a charming, self-sacrificing buddy (often the same sex, but ideally the opposite sex) whose sole role is to attract potential dates to you. Your wingman introduces people to you, helps you keep the conversation moving, smiles at you, puts you in the best light, and makes you the center of attention. His role is to remain as support only, in the wings, leaving center stage for you.

Does a guy love you less after looking at (other) beautiful women?

A friend ten years my senior once told me never to marry a photographer, a college professor, or any male in the hospitality, fashion, or entertainment business. Better to pursue an engineer or a construction worker. "Why?" I asked innocently. She gave me a serious look and said that men are influenced by what they see around them. If all they see all the time are young, attractive, doting babes, it's a problem. No matter how pretty you are, the constant comparisons don't work in your favor. It was just a worldlier woman's well-intentioned advice, but it may turn out to have a grain of truth to it. It's now known, for instance, that male high school teachers and professors have suspiciously high divorce rates, whereas male kindergarten teachers tend to be monogamous.

Psychologists Douglas Kenrick and Sara Gutierres at Arizona State University study what is known as the "contrast effect," which occurs when a person of average attractiveness is directly compared to gorgeous people and judged as less attractive than he or she would otherwise be. In one of the duo's studies, undergraduate men who recently watched the beauties starring in an episode of *Charlie's Angels* were significantly less willing to date average-looking women than were men who hadn't recently watched the show.

In a similar study, the psychologists recruited couples who lived together and were in committed long-term relationships.

Avoid the Contrast Effect

Comparisons are odious, as the saying goes. As much as you should avoid partners who constantly compare you to other beautiful babes, the contrast effect is even more dangerous when you do it to yourself. According to psychologist Jonathan Brown and his colleagues at the University of Washington, the characteristics of other people color our perceptions of ourselves. We're like Gulliver in Jonathan Swift's *Gulliver's Travels*, who becomes either a giant or a pygmy depending upon who's around him. Brown found that women's self-appraisals were less favorable when they viewed photos of stunning people of the same sex and more favorable when they viewed pictures of unattractive people. Moreover, those with low self-esteem were more likely to get a boost by identifying with someone who was exemplary in some way. Either way, it's a dangerous path. Be aware that your self-perceptions are the result of the contrast effect, and center yourself so you're confident in any context.

The experiment, ostensibly on art appreciation, required the participants to view pictures of either "well-proportioned" opposite-sex centerfolds or abstract art. After viewing either the hottie or the art, the men and women answered questions about their feelings for their partners. Strikingly, the men who had been gazing at the babe indicated less attraction to and love for their girlfriends or wives than those men who had been looking at abstract shapes and colors. Meanwhile, for women, looking at pictures of beefcakes was "pleasant" but didn't affect their feelings or desire for their partners. (However, in another study, women did downgrade their partners after reading about men who had status and resources.)

It's bad enough for a woman if her sweetie just looks at beautiful women, and it's worse if those women are flirts. In a similar experiment at McMaster University, researcher Sandeep Mishra and his colleagues showed men videos of an attractive woman. Some of the men watched the woman radiating warmth and implying the possibility of a future interaction, while others watched the woman acting neutral and uninterested. The group of guys who watched the woman acting flirtatious rated their own partners as significantly less attractive than did the men who saw the same beauty when she acted bland. (Ladies, remember this: looks alone don't always cut it.)

The contrast effect is completely subconscious. Men really don't mean to think you're less attractive or love you less after seeing young, hot chicks (even if you're also one). Mass media and everyday life introduce an abundance of gorgeous gals, and comparisons are inevitable. When the mating pool seems to expand to oceanic proportions, an average-looking woman, or even a pretty one, no longer seems like a good catch.

Exacerbating it all is men's knee-jerk neurochemical response to the sight of receptive, beautiful women. Men are known to react strongly and irrationally, thanks in part to high testosterone levels and activity in the amygdala, the part of the brain that makes spur-of-the-moment decisions. (See page 194 for more on what goes on in men's brains when they see sexy women and page 164 on the overperception bias that makes men think women are more interested in them than they are.) Some men raise their criteria and expectations unrealistically, while others soon recalibrate when they realize their mate value really isn't high enough to attract an Angelina Jolie or even the flirtatious bartender at the local pub.

Bottom line: my friend may be right about avoiding men who work around pretty young things all day. Just to be safe, I married a guy who spends most of his time on the phone.

Why shouldn't you spill everything about yourself on a first date?

I'm hopelessly nearsighted. Anything across a room from me appears hazy, yet I don't like to wear glasses. With perfect eyesight I see all my flaws and those in everyone and everything else, so it's best to go soft focus. Only when something is important enough to warrant further detail do I whip out the specs.

Many people subconsciously follow this same logic in a dating context, according to a study led by Michael Norton, a marketing professor at Harvard Business School. The reason why less is more, or blurry is better, is that people commonly prefer others who are similar to themselves (or at least how they *perceive* themselves). The more information you learn about someone you don't know well, the less you'll think you have in common, and the less you'll want to date that person. You may *think* you like people more when you know more about them, but this is not necessarily true, at least not in a dating context, and not in the very beginning of a relationship. Again, better to go soft focus.

Norton and his colleagues discovered these biases in a multi-trial study involving students on the MIT and Yale campuses and singles in their thirties who use Internet dating Web sites such as Match.com and eHarmony. In one trial involving several hundred participants, the researchers showed subjects between one and ten traits that described a potential date (bright, cultured, methodical, individualistic, polite, stubborn, and so on). The greater the numbers of traits revealed, the less the subjects thought they had in common with their prospective date, and the less enthusiasm they had for dating that person. In another trial, participants were shown traits one by one in random order. If the first trait people saw didn't apply to themselves, they were less likely to find further similarities and more likely to conclude the person wasn't compatible with them. If they identified with the first trait, they perceived themselves as having more in common with that person and were more tolerant when dissimilar traits came up later.

It's a bias toward the similar, especially for long-term relationships. We all seek patterns, and if it seems early on that you have something in common, or if the information you have is intriguingly vague, you'll find evidence to support a match. If you decide right away that you don't have much in common, you won't. Others do the same when they meet you. This means if you want them to like you, it helps to overemphasize your similarities, not your differences, from the get-go. (This explains the popularity of sites such as eHarmony and matchmaking services that claim to do all the personality matching for you. Your pool of candidates is presorted, the heavy lifting done, and you have a stronger conviction that the people you meet will have a lot in common with you.)

How fickle and fanciful relationships are in the very beginning! So easily do they tip one way or another. It's funny, but we often fall in love the easiest when our impression of a person is unburdened by too much information. The trick is to reveal just enough about yourself—and to extract just the right amount of information from your date—for both of you to learn more about each other yet reduce the risk of making a premature rejection. It's a careful balance of fantasy and reality—like seeing 20/20 through rose-colored lenses.

Revealing Less Is More

Upon meeting women online, my friend Seth used to disclose four things about himself: he dropped out of an Ivy League grad school, lost a load of money in a bad business deal, is a card-carrying libertarian, and sometimes gets road rage. He thought he was being transparent and descriptive, and that women go for that, but one prospective date flat out told him it was too much information. Think about it: do your dates really need to know about all your peeves and prejudices? If early dissimilarities create a negative bias, you might want to be a little ambiguous when you first get to know someone. Avoid summing yourself up with supercharged words or words that limit you, such as *political* or *methodical*. Don't spill all your negatives in the beginning. A Rutgers University study found that singles on Match.com who reported they were more honest and up-front about their negative traits had less success (although general self-disclosure was considered a good thing). Full disclosure can wait until the second date or later.

Introduce yourself in terms that allow open-ended interpretation. Describe what you'd like to contribute to a relationship, rather than the hard facts. Be humorous—it's a way of revealing something about yourself without going into the nitty-gritty. Find out what you have in common and emphasize it. And if you find yourself rejecting too many people and regretting it, ask your prospective dates funny or leading questions about themselves that will bias you toward liking them more before you pull out the magnifying lens.

Why do you overestimate your competition?

There are two mistakes you can make in trying to determine your mate value (that is, your desirability compared to other singles in the mating market): assume your competitors are more desirable than they actually are, or assume they're less desirable than they actually are. Both come with risks. If you underestimate—*She's not nearly as pretty or interesting as I am, so he'll never go for her*—you might be too cocky and find yourself unpleasantly surprised. On the other hand, if you overestimate your competition—*God, she's so gorgeous and so talented; what would he ever see in me?*—you might give up too easily and settle when you shouldn't. So what do you do?

If you're like most women, you overestimate; that is, you think other women are much more attractive to guys than guys really think they are, according to a study by evolutionary psychology researcher Sarah Hill at the University of Texas at Austin. Hill recruited nearly five hundred people to look at photos of men and women and rate their looks. She asked one group of guys to look at women's photos and judge how sexually desirable they are and how much they'd want to date them. Women also judged the ladies' looks and guessed how much they thought the men would find them attractive. A different group of men and women did the same for the photos of men. It turned out that women significantly overestimated the appeal of other women, and men overestimated the appeal of other men.

Do we all have self-confidence problems? Not exactly. According to Hill, it's more likely that women and men have evolved to err on the side of thinking others are more attractive and desirable than they really are because that's less risky than being overconfident. If you think other women are so great, you're more likely to try harder to make yourself more desirable, promote yourself more, work harder to keep your partner, value your relationships more, and not waste your time on people who are way out of your reach. This increases your outcome of getting a partner and keeping him. (The same goes for guys.) Sure, you might risk giving up on someone even better, but that's not as tragic as rejecting too many people or not being aware of a partner who cheats on you because you're too cocky.

Of course, there's an ugly side to overestimation. By thinking others are so desirable, you might act a little crazy. You've gone too far if you find yourself trying too hard to attract a guy, bending over backward more than you should, and coming across as too desperate. You need to rein in your overestimation bias if, in a relationship, you find yourself always jealous of other women your partner meets, and paralyzed by fear that he'll leave you for them. Men also go too far, by being overpossessive of you, tracking your every movement, and forbidding you from going to places where you'd meet other guys. However, once in a stable, committed relationship, couples may overcome the overestimation bias and become less competitive against members of the same sex. This is especially true when members in a couple consider their mate to be their equal (or lesser).

The key to overcoming uncertainty when dating is to end all the guesswork and try to get as accurate a picture as possible of your "mate value." The only way you can do this is by dating, dumping, and getting dumped—which means sometimes overestimating and underestimating, and sometimes getting it just right.

Know Your Overestimation Bias

In your heart it's useful to know that you probably have a tendency to overestimate your competition. Don't let it go to your head, but other women are probably not as desirable to men as you think they are. Let this knowledge give you the confidence to pursue someone who seems really worthwhile. It's also useful to know that men are often just as insecure, thinking that other guys you meet are more desirable to you than they really are.

It happens that the overestimation bias works outside of dating, too. Whether you're applying for a job or playing a game of chess, you'd do better to overestimate your opponents than underestimate them, provided you have a healthy enough self-esteem. By thinking the competition is stiffer than it is, you're more motivated and more prepared, and you reduce your risk of losing.

Is there an ideal number of people to date before you settle down?

Let's be theoretical. There are one hundred men who are aching to marry you. You have no idea how good any of these guys are until you get to know them. If you took the time to date each one, you'd need a cane to walk down the aisle on your wedding day. You're choosy yet mortal. So what do you do?

To solve this problem, Peter Todd, a cognitive scientist and psychologist at Indiana University, used a concept called *satisficing* (satisfying and sufficing). In this context, satisficing involves setting a standard and then searching for someone who not only meets it but also exceeds it. This means that even before you think of settling, you'd date a number of men just to establish a baseline. The pickier you are, the more men you'd want to meet before you set the standard. The problem is that you don't want to date too many guys beforehand, because you might accidentally dump the perfect guy (in theory, you can't go back to a guy after you dump him). So how do you maximize your chances of finding Prince Charming and minimize your chances of accidentally dumping him?

In Todd's mate-match model, if you're an extremely picky person who's dead-set on finding the best guy in your dating pool of one hundred guys, you'd set your baseline after dating 37 percent (37 out of 100) of them. After you've dated that many guys, you should feel fairly confident about settling for the next guy you meet who exceeds the standard. If you have more relaxed standards and care only about finding one of the best guys for you, you'd "satisfice" after dating only 9 percent (9 out of 100) of the men in your dating pool. The next guy you meet who's better than any of those in your sample gets to be your husband.

But in the real world guys get to choose, too, which complicates matters. Not every man wants to commit to you, and you need to deal with rivals who are snapping up the eligible bachelors. To address this cold reality, Todd and his colleague Jorge Simão, a computer scientist at the University of Lisbon, created an algorithm that simulates a social network of one hundred men and one hundred women. This time, like in the real world, you're allowed to hold on to a guy you're dating while simultaneously meeting others. As you flirt, date, and get dumped, you get a pretty good idea of your own "mate value," that is, how attractive you are to potential mates, or your rank in the singles pecking order. Once mate value is established, the matching process becomes efficient because you know not to waste too much time pursuing a Brad Pitt, and undesirable men figure out they shouldn't waste too much time on you. You can stay with a guy without committing to him and "upgrade" when you meet someone better, assuming he likes you, too. As in the real world, options diminish over time as others pair off and the pressure to commit increases.

At a certain point, you'll probably decide to "satisfice" for the man you're dating—which, according to the simulation, happens after dating somewhere between one and four men. (Remember, this is different from the previous mathematical models in that you get to check out other guys while holding on to the one you're with, and they get to do the same.) You may not get your first or fifteenth choice of the hundred men, but there's a good chance you'll end up with someone who's more or less your equal in mate value. Even as new people are introduced and as your peers marry and move off the scene, you'll probably fall within the 90 percent of couples who find partners. (According to cognitive scientists, 85 to 95 percent of the people in most populations in the real world are able to find a committed partner at least once in their lives by "satisficing" and/or holding on to the one they're with while looking around.)

As great as this sounds, with all the social, biological, and psychological variables of human attraction and commitment, it seems difficult, if not impossible, to create a simulation that perfectly matches reality. Maybe the reverse sounds better—matching the real world to a virtual one. That way, you really could have that kingdom of one hundred Prince Charmings all competing for you, and live happily ever after.

Applying Mathematical Models

If there are lessons to glean from mate choice mathematical models, it's that it's good for you to be picky and it's best not to settle until you've played the field a little, but don't spend your entire life searching for the one right person. If you're not sure if you want to commit yet, check out other guys even when you're dating. Another lesson is that it may be better to pursue a guy (even subtly) than wait for one to pick you up. In another famous mate-choice simulation, the Gale-Shapley Stable Marriage Algorithm, men make offers to women, but women cannot make offers to men, and women are allowed to decline an offer only in favor of another man who proposes to her. It turns out that guys get the better bargain. Men on average get to marry women who rank high on their wish lists, while women on average end up settling with men who rank low on theirs. Of course, the real world is more complex, but there's still a valid lesson here. Pick and be picky!

CHAPTER 9
Love on the Brain

And of course, the brain is not responsible for any of the sensations at all. The correct view is that the seat and source of sensation is the region of the heart.

—Aristotle, in *De partibus animalium*

How does being passionately in love change your brain?

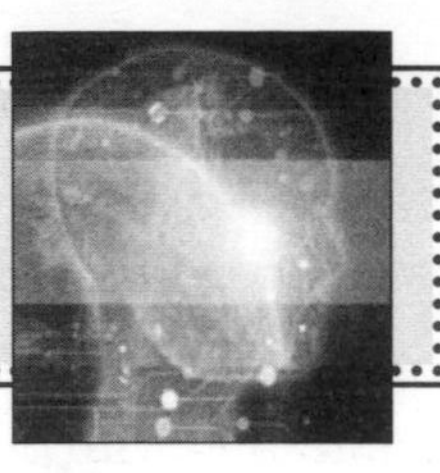

If you're lucky, you don't just fall in love; you fall *madly* in love. You can't eat, can't sleep, can't focus on anything else. You're living, eating, dreaming, and breathing for love. Love "rewires" your mind, throwing switches that send shivery charges and hormonal surges. Your brain on love resembles the brain of a person on drugs or gone temporarily insane. It's a manic, driven, addictive brain. We know this because researchers have used magnetic functional resonance imaging (fMRI) to explore just how mad love can look.

Helen Fisher, an evolutionary anthropologist at Rutgers University, Lucy Brown, a neuroscientist at Albert Einstein Medical College, and Arthur Aron, a psychologist at Stony Brook University, have done some of the best-known research on the romantic brain. Several years ago, the trio recruited men and women who were intensely in love. They had been with their partners for around a year or less, which means they were still in the first, passionate stages of their relationships. They were slid into an fMRI scanner, one by one, and told to gaze at a photo of their darling.

Right away on the brain scans, the researchers saw a glow in several regions of the lovers' brains, representing increased blood flow. Among them were three important clusters of brain cells related to motivation and reward. A big one called the right ventral tegmental area (VTA) produces a feel-good neurotransmitter called dopamine. Another cluster, the medial caudate nucleus, plays a role in integrating learning, memory, pleasure, and communicating with other goal-oriented parts of the brain. The third, the nucleus accumbens, is a reward center that is also activated when a person is high on cocaine or another addictive substance. The three clusters are connected like lights on the same circuit.

In those magical early stages of passionate love, the VTA releases dopamine to the caudate nucleus and nucleus accumbens. Dopamine makes you feel great when it tickles these target areas. It gives you superhuman energy. You're carried away in a sea of euphoria, desire, even mania. In your most passionate moments the hormone norepinephrine (adrenaline) also rushes over you, which, in concert with your sympathetic nervous system, makes your heart pound, your knees tremble, and your palms sweat. Adrenaline also heightens attention and short-term memory and is associated with goal-oriented behavior.

When this neurochemical love potion is strong, you and your sweetie are up at all hours, lusting and longing and touching. You're insatiable. Your caudate nucleus and nucleus accumbens signal the VTA to send more dopamine their way. It does, and then it sends more and more as your love addiction intensifies. The researchers point out that these regions of the brain aren't emotion centers—they're *reward* centers. For men and women alike, love is a motivation and even an addiction. Some in this state might say it's the only reason for living.

When you're in love, regions of the brain associated with sexual motivation may also be activated. Dopamine triggers the production of the hormone testosterone in women, which amps up the sex drive and may make you more aggressive. Intriguingly, men's levels actually drop when they fall deeply in love, which explains why many guys become less aggressive and confrontational once they have partners. As your true love caresses you, you're also hit with the pair-binding hormone oxytocin, a neurotransmitter released from the pituitary gland that promotes trust and attachment to a specific person. Oxytocin levels rise when you're cuddling and orgasming; the hormone affects both sexes but is more potent in women (see page 144). For men, vasopressin, a similar neurotransmitter, is also released when falling in love or having sex, and is linked to mate bonding and territoriality (see page 304).

A team of Italian neuroscientists also found evidence that in the first six months of a passionate love affair, men and women have higher levels of nerve growth factor in their blood. The more intense the infatuation, the more brain protein they produced. It's unclear what nerve growth factor does to you psychologically, but it is known to regulate the growth and survival of neurons throughout the nervous system, including the heart and brain, and also modulate hormones. Of particular interest, nerve growth hormone may increase levels of vasopressin, which is crucial to men's ability to bond with their partners.

While some parts of your brain are activated in early love, others basically shut down. Taking a backseat are the prefrontal cortex, which is responsible for social judgment and critical faculties, and the amygdala, the part of the brain that triggers fear. The deactivation of the prefrontal cortex and amygdala means you more easily overcome any feelings of distrust of your darling. These regions of the brain are associated with anxiety and "overanalyzing." For lovers, judgment is suspended. Deactivation of these faculties, in concert with everything else going on in your brain, also helps explain why you do crazy, impulsive things when you're in love, such as drive through an ice storm just to see him, or fly to Vegas to get married by an Elvis impersonator.

And if that's not enough, your brain in love absorbs up to 40 percent less of the neurotransmitter serotonin, and low serotonin levels result in obsessive-compulsive thoughts. You have no appetite for food or sleep. You spend hours Googling your new love, his family, his friends, his kindergarten classmates. You fixate on memories: his gestures, the way he grasped your wrist in his hand, that look he has when he's about to kiss you, the nuanced comment he made about you being the one, and so on, ad nauseam. (I remember obsessively checking my e-mail to see if my new man, an imaginative writer, had written me. And when I was away from my computer—at the gym, doing work, riding the subway—all I could think of was the sweet cryptic content of his last love letter and the prospect of the next.) Low serotonin, which results in obsessive thoughts, is the reason why many love-struck people claim to think about their lover 95 percent of the time (which adds up to a lot if you're also not sleeping).

The researchers noted some preliminary brain differences between men and women in love. Passionate guys showed more activity in the visual cortex, in keeping with the theory that men are wired to pick up visual cues, and in parts of the brain involved in penile erections. Meanwhile, women in love showed more activity in the parts of the brain such as the posterior parietal cortex, which is somewhat of a mystery region but thought to be related to attention and consciousness. The hippocampus, which plays a role in long-term memory, also glows hot in women in love but not in men. This may explain why you vividly remember all the details of your date—what dish you both ordered, the waiter's snarky comments, the conversation and accompanying expressions, the color of the carpet, and so on. Your guy, who doesn't remember these details, had less going on in his hippocampus.

Like any addictive drug, the effects of early love gradually wear off (usually within a year), and the chemical tsunami in your brain becomes a gently lapping pool, subject only to occasional storms. Our species wouldn't survive if love's rapture lasted any longer than it does. You'd have no food, no shelter, no clothing, no career—only your partner. Which sounds great only when you're truly, madly, deeply in the first stage of love.

Reconsider Antidepressants

The most popular antidepressants (such as Prozac, Zoloft, Paxil, and Lexapro) are selective serotonin reuptake inhibitors (SSRIs). Serotonin is a neurotransmitter that controls impulses and obsessive behaviors and gives you the feeling that you're calm and in control. Sounds like you'd want to have a lot of it, right? The problem is that the obsessive, intrusive, disturbing, arousing, and irresistibly pleasurable dwelling on the details of your love affair is possible only when serotonin levels are low and levels of the stimulant dopamine are high. This means that taking serotonin-boosting SSRIs isn't very good for falling in love, according to a paper by Helen Fisher, an anthropologist at Rutgers, and psychiatrist J. Andrew Thompson.

According to Fisher and Thompson, if you're taking SSRIs, you might not pick up on the usual subtleties of attraction or send out sexy signals to others. The drugs may hinder your instinctive ability to tell if someone is bad for you, because your anxiety levels plummet as your serotonin levels rise. Antidepressants may also reduce your ability to have an orgasm or even get sexually aroused, which in turn inhibits your ability to bond with your partner. In short, antidepressants can be anti-passion.

Despite concerns about antidepressants' depressing effect on love, going off them is obviously not the answer for people suffering from serious depression. The researchers' warning applies only to psychiatrists who overprescribe them to people who may otherwise find an even better cure—such as love.

Why do you find your partner so amazing, even if no one else does?

"Love is a gross exaggeration of the difference between one person and everybody else," said George Bernard Shaw, and the statistics back him. Ninety-five percent of us think our partners are above average in appearance, intelligence, warmth, and sense of humor. With so many exceptional lovers, it's a wonder that any of us ever break up.

Commitment to a long-term relationship requires a conviction that your darling is better for you than anyone else, and such sureness requires a healthy dose of self-delusion. If you didn't idealize him (and if he didn't idealize you), there would be no ideal, and without an ideal the future seems bleak, doesn't it? The more idealized your perception of your beloved, the more satisfied you probably are in your relationship.

This is the gist of what psychologists call romantic idealization. A study of nearly two hundred married or dating couples led by psychologist Sandra Murray found that a predictor of happiness in a relationship is whether people idealize their partners. Generally, people were happiest when they idealized their partners and their partners idealized them, even if both parties didn't see the same virtues in themselves. Whether married or dating, when men and women saw their partners in the imperfect and inadequate way their partners saw themselves, not in a rose-colored light, satisfaction ratings took a turn for the worse. Men in particular were happiest when their partners idealized them, regardless of whether the perception was warranted. Flattery apparently doesn't hurt a relationship.

Love-illusioned couples are also happier than "realists" in the face of the inevitable adversities that afflict every relationship. They feel more secure, empowered with the conviction that they are with the right person, even if he's unable to hold a job or she's hopelessly neurotic. When good things happen, the partner is credited, and when bad things happen, the situation is to blame. In this rosy glow, she perceives him as a brilliant nonconformist and he perceives her as poetic and sensitive. Of course, idealizing without idolizing is key. It's not as though you need to think he's great all the time at everything. The researchers found that the happiest couples cast their disappointments in the best light without denying negativity.

Bear in mind that romantic idealization happens without deliberate self-deceptions. Your brain has brainwashed you. Using fMRI, neuroscientists have found that when people who are deeply in love think of their mates, regions of the brain involved in social judgment and negative emotions, the prefrontal cortex and the amygdala, are switched off. With the amygdala and prefrontal cortex silent, you may not objectively see your sweetie's faults or distrust him. (This only applies to your partner; you remain perfectly capable of being critical of anything else you encounter. You might even be especially critical of those who criticize him.) At the same time, the "reward" areas of your brain, the caudate nucleus and the ventral tegmental area, are switched on, encouraging you to keep the relationship going. And when you kiss and cuddle and make love, you're hit with oxytocin, which calms and soothes your nervous system.

Bottom line: if you think your lover is the most amazing creature on earth, that's fine and natural—and there's no reason to care if others don't see the same. And if he thinks the same of you (this part is crucial), then he probably *is* the one for you. It's likely to be a self-fulfilling prophecy, and isn't that wonderful? There are very few things for which it's okay to have idealized, self-deluded conviction. Let love be one of them.

> *To be in love is merely to be in a state of perceptual anesthesia—to mistake an ordinary young man for a Greek god, or an ordinary young woman for a goddess.*
>
> —H. L. Mencken

Why do couples look increasingly alike over time?

It's likely that in a long, loving relationship, you and your partner will look like each other. I don't necessarily mean that you've picked someone who looks like you, although this happens, too. What I mean is that over time, your faces resemble each other's more and more. Yes, as the years pass your features might *converge*. How scary is that?

In a landmark study on the appearance of married couples, Robert Zajonc, a psychologist at Stanford University, and his colleagues asked twelve married couples to send two sets of individual portraits of themselves, one set taken in their first year of marriage (newlywed photos) and the other after twenty-five years of marriage (old-timer photos). The researchers put the newlywed photos together in one pool and the old-timer photos in another, and recruited nearly eighty participants to guess which men and women were married and look alike. It turned out that matching the newlywed couples was impossible—the raters were no more accurate than they'd be by dumb chance (the couples were all the same race, ethnicity, economic class, and approximate age). But when matching the old-timer photos, raters were shockingly spot-on. That's because couples look more like each other on their silver wedding anniversaries than they do as newlyweds.

Zajonc and his colleagues drew on emotional efference theory to explain what happens: in short, people who empathize with each other mimic each other's facial expressions. Mimicry is an unconscious and involuntary process. When you mimic another person's facial expressions, you subjectively feel that emotion or mood. (See page 179 on mirror neurons.) Over time, with use, facial muscles either grow or atrophy, just like your biceps or calves. When facial expressions are repeated and habitual, like a ready smile or a constant grimace, they permanently etch a "look" into the face. Furrowed brows, puckered lips, stress lines, laugh lines, crow's-feet, and lines between the eyes or around the mouth happen over the decades, just as running water contours rock.

The happier the marriage, the more spouses grow to resemble each other. Zajonc surveyed the twelve couples in the study, asking them questions about their satisfaction in the relationship and the incidence of very happy or tragic experiences they had together over the decades. He found that the more they shared attitudes, the greater their mutual resemblance. That's because they'd been laughing and crying and worrying together for so many years.

Other factors may also contribute to facial convergence. Diet is one, although the husbands and wives in the study did not resemble each other in their quantity of facial fatty tissue. Climate is another—living together in a hot, sunny place may weather couples' faces similarly. Moreover, people with similar temperaments may also develop similar expressions. Seen this way, like attracts like, and like becomes even more alike over time.

The best advice you can get when about to marry is to make sure you love your partner—and the way he looks when he expresses himself, because that's your future.

Help Your Sweetie Smile

If your partner is depressed, get help. It will help you both psychologically—and preserve your beauty. For several years my friend Ada dated a man who was clinically depressed. How easily she empathized with his unhappiness and frustration. She wasn't conscious of it at the time—we rarely are aware of our habitual expressions—but she looked depressed. Only after the breakup did Ada's friends tell her how much happier and younger she appeared, and how miserable and lined her face had looked for so long. On reflection, she realized that she was no longer mirroring her partner's sour face. Even over a relatively short period of time we may take on the look of the people we love.

Does being in love make you blind to other people's love?

Psychologists have long noted that the average person is uncannily accurate when making snap judgments based on very little information. From video clips that are thirty seconds or less, impartial observers can often correctly tell whether two people are friends or rivals, gay or straight, whether the rapport is good, and whether the people are extroverted, conscientious, and have high testosterone levels. But what about love—does everyone perceive love as accurately?

Psychologist Frank Bernieri and researcher Maya Aloni thought that people in love would perceive love more accurately than those not in love—and they were in for a surprise. In a study set at the University of Toledo, they asked 150 men and women to watch twenty-five-second video clips of couples who were in various relationship stages—from just friends to madly and passionately in love. The participants, themselves in relationships with varying levels of commitment (or not in one at all), were asked to judge if the couples in the videos were in love, and if so, how much in love. It wasn't obvious—the couples on tape were having pleasant discussions about their lifestyles and what they enjoy doing, but offered no direct clues on their level of love and commitment.

Among the raters, the people most in love were the most confident about their ability to identify others in love. Astonishing to Bernieri and Aloni, they were also the least accurate. This was especially true of guys. Men in love were significantly worse than other men in identifying love in others. Women in love were also less accurate than other women, although not significantly so. Perhaps men are particularly bad at identifying love because they don't talk about it as much as women do, so they don't recognize it in all its incarnations.

The lesson is that love *is* really blind; that is, to all love but one's own. There's a blind spot in the brain when it comes to the critical judgment of love. Lovers often become fixated on their own love life and use it as a template for the way love should look. If holding hands, looking into each other's eyes, and other physical displays of affection are a big part of your relationship, those displays might be the only way you perceive love. It doesn't matter that the way you express your love is not the way other couples express theirs. Perhaps it's arrogant or self-centered of you, but you try to force all love to look like your own, at least subconsciously.

I think there is another reason why singles are more accurate than lovers at detecting love. When you're single, it might be in your best interest to be more perceptive about the state of other people's relationships. Is that couple really in love, or do you have a chance with her hot boyfriend? It's often said that love is blind, but jealousy has 20-20 vision.

Are people naturally monogamous?

Yes and no; it depends on how you look at it. As actor Kate Hudson delicately explained, "I don't think monogamy is realistic. But I believe that we, as a people, have the power to make it happen." Humans are among the 4 percent of mammals who have exclusive sexual relationships, which is one definition of monogamy. But the rule is interpreted loosely, as we all know. Most of us don't stay with one partner for our entire lives. Instead, we practice "serial monogamy" by having many sequential sexual relationships over a lifetime. We might flirt with and fantasize about others when in a monogamous relationship. Both sexes are known to have affairs. Even short-term monogamy isn't for everyone.

Unconvinced that our species is monogamous? Then think of it this way: we're only *mildly* polygamous. Mildly, because when you compare us to other species we're downright puritanical. Purely promiscuous mammals reveal at least one of two telltale signs in their anatomy. In species in which dominant males cultivate harems of females, there is a tremendous difference in body size between the sexes. The male gorilla, for example, needs to compete against other males for sexual access, so he's up to twice the size of the female, which is certainly not true of humans. The second sign is scrotal size. In a species in which both males and females are promiscuous, males have larger testes. The bigger the testes, the more sperm is produced, and the better chance a fellow has of flooding out his competitors. A typical swinging bonobo male, who has sex dozens of times a day with different partners, bears a scrotal sac as big as a grapefruit, whereas a human male carries a modest sac that varies between walnut- and orange-sized.

As our ancestors evolved over the last couple of million years, infants' brains became bigger and their dependency on parents lasted longer. Females needed extra calories for pregnancy and lactation, and to help raise an infant. These pressures might have spurred a need for a dedicated mate (at least for a while). Concealed ovulation may help empower females with the right to choose and retain partners (see pages 128–129 for the reasons). Some evolutionary biologists suggest that our remote ancestors *Homo ergaster* started to couple up (pair-bond) sometime after they emerged as a species, around 1.7 million years ago. The details and duration of these early human love affairs are unknown.

While mothers benefit from having a man around, the arrangement also benefits men. Although it's often said that men spread their seed while women harvest it in one place, fatherhood may make men decide to plant a garden rather than a forest. (See page 148 on testosterone levels of fathers.) By spending exclusive time with one woman, a man maximizes the chances that he's the father of her children and that his support goes to raising kids who are biologically related to him. Children whose fathers helped raise them probably had a better chance at survival.

Observing the diversity of genes in human populations, geneticists have found evidence that strict monogamy was never the norm for us, but it became more common between five thousand and ten thousand years ago, when cultures transitioned to sedentary farming communities with households. This shift meant more men got to marry, whereas in hunter-gatherer societies that were polygamous (not all were), high-status men had several wives and many children while many lower-status men didn't have any. In a society without a distribution of wealth and resources, many guys wouldn't get to marry and have children. (Tell that to your partner next time he complains you're keeping him on too tight a leash.)

Strict lifelong exclusivity is not common in Western countries.

Half of all marriages in the United States end in divorce. Among married American couples, 22 percent of men and 14 percent of women admit to having dallied at least once, and some studies indicate much higher figures. Even with birth control available, at least 3–4 percent of births are the result of illicit affairs. Infidelity rates in other cultures that are tolerant of it, and even romanticize it, are obviously higher. France's first lady, Carla Bruni, who has been a supermodel and singer, once said, "Monogamy bores me terribly. . . . I prefer polygamy and polyandry."

Looking at the monogamy question through an evolutionary lens, Rutgers anthropologist Helen Fisher suggests that ancestral human relationships may have had four-year cycles—just long enough for a couple to raise a baby past infancy (others have cited a "seven-year itch"). Parents go their separate ways, eventually forming a new family, with older kids helping to raise younger ones.

The bottom line is that for most people monogamy doesn't come naturally, especially if defined in the stringent sense of one mate for life. Culture and personal experience are and always have been major influences here. The spectrum ranges from true monogamy to extreme promiscuity (which is still pretty tame when compared to our wild mammalian cousins), with most people in Western countries practicing something like serial monogamy. Where you and your partner fall depends on the usual blend of personal choice, genes, background, and culture. Let's hope you match up.

Why does absence make men's hearts grow fonder?

Absence makes your partner's heart grow fonder—if you measure fondness by how amorous he is after spending time apart from you. Not only is he fonder, he probably thinks you're more attractive than usual, and believes other men think so, too. He also is more easily aroused, has a stronger urge to have sex with you, and holds a stubborn conviction that you want to have sex with him, too. If it's not the heart that grows fonder, it's the nether region.

Your partner's ardor after not seeing you is not simply due to time apart without sex. If the same period of abstinence were spent in each other's company, he wouldn't be so passionate. It's uncertainty that drives a man to act this way, according to evolutionary psychologist Todd Shackelford and his colleagues at Florida Atlantic University, who surveyed four hundred young men in committed relationships and arrived at these results. Unconsciously, your partner may fear you were unfaithful. The more time apart, the greater the possibility that you were cavorting with another man.

Seen from a purely genetic perspective, it's a bad deal for a man if his time and resources go into raising another guy's kid. That means his instinct, honed over centuries of evolution, combined with a culture of machismo, is to respond with agitated passion to the prospect of you having sex with another man. As a

Watch the "Mate Guard"

Guys who think their partners are unfaithful tend to be either vigilant or violent, according to a study led by evolutionary psychologist David Buss. The better-looking you are, the more likely your partner is to "guard" you. He may hide you from rivals and try to prevent you from meeting guys who are wealthier and better-looking. He might call you at unexpected times to see whom you're with, what you're doing, and where he can find you. In public, he may throw his coat over your shoulders or place a possessive arm on your body. He may even pick a fight with other guys who dare to glance at you, especially if he's tormented by "Othello syndrome," the belief that one's mate, or the woman a man wants as his mate, is so desirable that every man wants to sleep with her. Remarkably, men have been found to instinctively guard their partners the most when women are near ovulation.

Women also get possessive of their partners, especially if those partners have high status or ambition. Motivated girlfriends and wives might doll themselves up or get liposuction to make sure they're competitive in the looks department. In public they stick closely by their partner's side, subtly alerting other women that he's theirs by touching him or caressing him publicly. They may express indignation or distress if he flirts with other women, helpfully pointing out their addiction to diet pills or their bad taste. If they're sly, they may even flirt with other men, on the assumption that blinding a guy with jealousy means that he won't notice other women.

result, he might push to make love as soon as possible after he sees you, so his sperm would compete against your lover's. If you were to get pregnant, he would have a better shot at being the baby's father. (Bear in mind, this drive evolved in men well before birth control, paternity testing, and child support.)

Amazingly, the higher the risk that a woman has been unfaithful, the more sperm a man ejaculates the next time he has sex with her, according to biologists Robin Baker and Mark Bellis. They found that, on average, a man who's been around his partner 100 percent of the time has a sperm count of about 350 million, compared to 800 million from a guy who's only been around his partner 5 percent of the time, regardless of how much time has passed since they've last had sex. Another study found that when men masturbate to photos of a woman having sex with more than one man at a time, they have a significantly higher percentage of motile sperm in their ejaculate.

The frequency of racy "cheating wives" scenarios in porn is a testament to the sperm competition theory. A married couple I know role-plays scenarios in which the wife cheats on her husband with a whole cast of imaginary characters, from the brawny sailors who roam the streets of New York City during Fleet Week to the liver-spotted homeless man to her plotting boss. Nothing makes my friend's husband a more passionate and aggressive lover than the idea that she recently romped with an unwholesome stranger.

Adding to the evidence, men who accuse their women of infidelity are more passionate—that is, they thrust deeper and harder during sex (according to the women). The reason, according to evolutionary psychologist Gordon Gallup Jr., who led the study, is that forceful, aggressive sex is men's unconscious attempt to displace a rival's sperm. Experimenting with simulated semen, dildos, and synthetic vaginas of many shapes and sizes, Gallup and his colleagues at SUNY Albany discovered that deep, vigorous penetration may remove at least 80 percent of the semen near a woman's cervix. The head of the penis is shaped like a scoop, and when a man thrusts, sperm accumulates behind the coronal ridge between the head and shaft of his penis. By removing a rival's sperm, a pregnancy is averted. The shape of the penis and its ability to expel sperm is also a reason why having sex for the second time in one night is a sloppy experience.

Knowing this, what can you assume the next time your partner shows up at your door after a long absence, bearing flowers and compliments? It might be the threat of sperm competition fueling his passion. For some women, raw ardor is a big turn-on. But there's also a sweeter alternative—it could just be that he loves you and missed you.

Do your genes make you more faithful or adulterous?

Wouldn't it be great if, along with the usual battery of STD tests taken before entering an exclusive relationship, your partner also took a monogamy test for the "cheating gene"? Disappointingly for those who would like science to provide clear answers, there is no such test, no single gene. Instead, many genetic as well as cultural and personal factors contribute to how faithful you and your partner are to each other. Naming them, you could go on for a while: parental role models, partner choice, opportunity to cheat, past loves, life stage, life satisfaction, religion, hormone levels, and so on.

Looking at genes alone, we know already that women's MHC genes can predict infidelity to an extent. The more overlap between a couple's respective gene variants, the greater the chance that the woman will have an affair, or at least fantasize about one (page 35). We also know that husbands with high testosterone levels have higher infidelity rates than husbands with lower levels of the hormone (page 150).

In recent years, scientists have had new insights on genes that influence male monogamy in particular. Neuroscientists Miranda Lim, Larry Young, and their colleagues at Emory University have found something of a "monogamy gene" in prairie voles, a species of rodent that mate for life. A male prairie vole spends most of his earthly hours grooming his mate and building their nest, and strays very rarely. Neuroscientists attribute the vole's fierce dedication to a feature of his brain anatomy: vasopressin receptors located in the ventral pallidum, one of the brain's pleasure centers. Vasopressin is a hormone that influences the activity of the feel-good neurotransmitter dopamine in the brain's reward circuits. (Female prairie voles, like female humans, respond more to the related hormone oxytocin.)

For a male prairie vole, lifelong "love" is very clearly due to his gene for vasopressin receptors. When the vasopressin receptors of male prairie voles are blocked and dopamine doesn't make it to the brain's pleasure centers, the rodents become cold and noncommittal. When the receptors are not blocked, males are faithful to the female they're with because they associate her with pleasure. When the researchers introduced the prairie vole vasopressin receptors in the brains of their promiscuous cousins, the montane vole, the playboy rodents became faithful homebodies.

Vasopressin also helps human males bond with their mates. Strikingly, men, like prairie voles, also show a high degree of diversity in their vasopressin receptor genes. It's possible that these genes could have some influence on men's ability to form relationships, although human love is clearly more complicated. Some researchers have already found evidence that the human vasopressin receptor gene may influence the ability to empathize with others. One brain study found that people in relationships for more than two years show increased activity in the reward area of the ventral pallidum, which is rich with vasopressin receptors. Do the men with active ventral pallidums also have the right type of vasopressin receptors there? It's possible. Is it possible that someday we'll have a gene therapy to deliver the right vasopressin genes to the reward areas of the brain, making men more faithful and happy in their relationships? Maybe, although there's surely more to it. (I have a paranoid male friend who thinks of it more in science fiction terms: a bioengineered virus that infects the vasopressin receptors of cheating men and makes them loyal, gormless househusbands.)

Oxytocin is another "pair-bonding" hormone (see page 144). There is some evidence that oxytocin receptors modulate attachment in female prairie voles in the same way that vasopressin receptors do in males. Oxytocin promotes bonding, and scientists have found that when oxytocin receptors in female voles' pleasure regions of the brain are blocked, the voles are not as attached to their mates. When oxytocin is injected into the brains of female prairie voles, they swiftly bond with the male who's with them at the time. As with female voles, women need oxytocin to form an attachment to their partners, and the amount of oxytocin absorbed by a woman's oxytocin receptors depends on her unique chemistry of genes and hormones. Do some of us have more, or at least more receptive, receptors? Possibly. Would women want gene therapy that helps them bond to a man? If so, can you really call it love? It gets complicated.

At this time in the field of neuroscience, trying to tell what's going on in people's minds from brain scans and neurochemicals alone is like watching the global markets from a satellite orbiting the earth. We can infer patterns judging from boats and buildings and other signs of activity, but we just don't know the details. Someday our knowledge of brain chemistry, genes, and hormones might be sufficient to understand why some people fall in love and are faithful for life . . . and others are cheating rats.

How does your brain "grow" when you're in love?

> *Love is the expression of two natures in such a fashion that each includes the other, each is enriched by the other.*
>
> —Felix Adler

According to cognitive scientist Douglas Hofstadter, in his book *I Am a Strange Loop*, two people in love internalize each other and create a shared state of mind that overlaps and expands their individual personalities. In essence, you form a "we"—a merger of your partner's attitudes, tastes, habits, experiences, knowledge, goals, and dreams with your own. You can think of this identity, the "we," as a pattern of neurons in your brain. It's a mental representation of your relationship. That's one reason why it's so darn painful to break up. You're "killing" the pattern—the "we" and the neural network of associations and rewards that supports it. If you've ever gone through a hard separation, you know it feels like murder. It is, in a way.

Neuroscientists suggest that a part of the brain that's key to creating the mental representation of a relationship is the angular gyrus, in the parietal lobe. Think of the angular gyrus as a bridge between your brain and the outside world. This bundle of neurons helps you incorporate new sensory experience in the context of your known experience. It enables you to make abstract representations of concepts, objects, and ideas. Without your angular gyrus, you wouldn't be able to step outside your mental state enough to be aware of it. You'd lose your ability to understand metaphors and abstraction ("reaching for the stars" would overwhelm you) or to read words and translate them into the internal language of your brain. Given its use in internalizing the external, the angular gyrus is thought to play an important role in love.

The angular gyrus is one focus of a study by neuroscientists Stephanie Ortigue of the University of California at Santa Barbara, and her colleagues Scott Grafton at Dartmouth and Francesco Bianchi-Demicheli at Geneva University Hospital. The team recruited nearly forty women who were happily in love with their long-term boyfriends or husbands, and put them in fMRI machines to track their brain activity during a word-detection test. (Only women were tested.) Before each stimulus, women were exposed to a subliminal flash of either their partner's name or a platonic friend's name. The flash of the partner's name made a significant difference in brain activity, which in turn affected women's test scores.

When the researchers looked at the fMRI scans of the women's brain activity after the partner's name flashed, they saw that the angular gyrus was active, along with the insula, which helps process emotional experiences, and the regions of the brain that are associated with reward and motivation, the caudate and ventral tegmental area (VTA). The researchers suggest that the sight of the partner's written name, translated by the angular gyrus, may have called up a mental representation of the relationship and triggered higher thought processes and associations involved in being passionately in love. At the same time, the emotional and motivational circuits of the brain became active. Only a subliminal flash of a partner's name, not a friend's name, stimulated these brain areas. When these regions lit up simultaneously as a "love-related network," women thought faster and performed better on the test. Interestingly, the more passionate a woman was about her partner, the higher she scored. Love, it seems, is a great motivator.

The neuroscience of love is still in its early, speculative stages. No one knows if being in love actually helps your brain perform better in everyday life, although it does appear to have a significant effect on the way lovers process information. Love may stimulate powerful unconscious associations that affect the way you perceive and respond to the world. It may motivate you in unexpected ways. Think of love as self-expansion, an internalization of the external. By association, you and your partner, the "we" you share and the world beyond, are all part of a network that grows with your love.

SOURCES

INTRODUCTION

Buss, D. M., Shackelford, T. K. (2008). Attractive women want it all: Good genes, economic investment, parenting proclivities, and emotional commitment. *Evol Psychol, 6*(1), 134–46.

Eastwick, P. W., & Finkel, E. J. (2008). Sex differences in mate preference revisited: Do people know what they initially desire in a romantic partner? *J Pers Soc Psychol, 94*(2), 245–64.

CHAPTER 1 FACE FIRST

Why do people seem more attractive when you are gazing into their eyes?

Rubin, Z. (1970). Measurement of romantic love. *J Pers Soc Psychol, 16*(2), 265–73.

Kellerman, J. L., Lewis, J., & Laird, J. D. (1989). Looking and loving: The effects of mutual gaze on feelings of romantic love. *J Res Pers, 23*, 145–61.

Williams, G. P., & Kleinke, C. L. (1993). Effects of mutual gaze and touch on attraction, mood, and cardiovascular activity. *J Res Pers, 27*(2), 170–83.

Aron, A., Melinant, E., et al. (1997). The experimental generation of interpersonal closeness: A procedure and some preliminary findings. *Pers Soc Psychol Bull, 23*(4), 363–72.

Adams, R. B., Jr., & Kleck, R. E. (2003). Perceived gaze direction and the processing of facial displays of emotion. *Psychol Sci, 14*(6), 644–47.

Adams, R. B., Jr., & Kleck, R. E. (2005). Effects of direct and averted gaze on the perception of facially communicated emotion. *Emotion, 5*(1), 3–11.

Mason, M. F., Tatkow, E. P., & Macrae, C. N. (2005). The look of love: Gaze shifts and person perception. *Psychol Sci, 16*(3), 236–39.

Berman, L. (2007). Looking for love? Eyes have it. Jan 22. *Chicago Sun-Times*.

Conway, C. A., Jones, B. C., DeBruine, L. M., & Little, A. C. (2008). Evidence for adaptive design in human gaze preference. *Proc Biol Sci, 275*(1630), 63–69.

Why do men prefer big pupils?

Hess, E. H., & Polt, J. M. (1960). Pupil size as related to interest value of visual stimuli. *Science, 132*, 349–50.

Hess, E. H., Seltzer, A. L., & Shlien, J. M. (1965). Pupil response of hetero- and homosexual males to pictures of men and women: A pilot study. *J Abnorm Psychol, 70*, 165–68.

Tombs, S., & Silverman, I. (2004). Pupillometry. *Evol Hum Behav, 25*, 221–28.

Laeng, B., & Falkenberg, L. (2007). Women's pupillary responses to sexually significant others during the hormonal cycle. *Horm Behav, 52*(4), 520–30.

What makes a face good-looking?

Langlois, J. (1990). Attractive faces are only average. *Psychol Sci, 1*, 115–21.

Perrett, D. I., Lee, K. J., Penton-Voak, I., Rowland, D., Yoshikawa, S., Burt, D. M., et al. (1998). Effects of sexual dimorphism on facial attractiveness. *Nature, 394*(6696), 884–87.

Slater, A., Von der Schulenberg, C., Brown, E., et al. (1998). Newborn infants prefer attractive faces. *Infant Behav Dev, 21*(2), 345–54.

Penton-Voak, I. S., Jones, B. C., Little, A. C., Baker, S., Tiddeman, B., Burt, D. M., et al. (2001). Symmetry, sexual dimorphism in facial proportions and male facial attractiveness. *Proc Biol Sci, 268*(1476), 1617–23.

Rhodes, G., Lee, K., Palermo, R., et al. (2005). Attractiveness of own-race, other-race, and mixed-race faces. *Perception, 34*, 319–40.

Rhodes, G. (2006). The evolutionary psychology of facial beauty. *Annu Rev Psychol, 57*, 199–226.

Thornhill, R., Gangestad, S. W., & Comer, R. (2006). Facial sexual dimorphism, developmental stability, and susceptibility to disease in men and women. *Evol Hum Behav, 27*(2), 131–44.

How long does it take to decide if a person is hot?

Kanwisher, N., Stanley, D., & Harris, A. (1999). The fusiform face area is selective for faces not animals. *Neuroreport, 10*(1), 183–87.

Olson, I. R., & Marshuetz, C. (2005). Facial attractiveness is appraised in a glance. *Emotion, 5*(4), 498–502.

Kanwisher, N., & Yovel, G. (2006). The fusiform face area: A cortical region specialized for the perception of faces. *Philos Trans R Soc Lond B Biol Sci, 361*(1476), 2109–28.

Kniffin, K. M., & Wilson, D. S. (2004). The effect of nonphysical traits of the perception of physical attractiveness. *Evol Hum Behav, 25*, 88–101.

Swami, V., Greven, C., & Furnham, A. (2007). More than just skin-deep? A pilot study integrating physical and non-physical factors in the perception of physical attractiveness. *Perso Individ Dif, 42*, 563–72.

Cloutier, J., Heatherton, T. F., Whalen, P. J., & Kelley, W. M. (2008). Are attractive people rewarding? Sex differences in the neural substrates of facial attractiveness. *J Cogn Neurosci, 20*, 941–51.

Are you more attracted to people who look like you?

Jedlicka, D. (1984). A test of the psychoanalytic theory of mate selection. *J Soc Psychol, 112*, 295–99.

Bereczkei, T., Gyuris, P., & Weisfeld, G. E. (2004). Sexual imprinting in human mate choice. *Proc Biol Sci, 271*(1544), 1129–34.

Wiszewska, A., Pawlowski, B., & Boothroyd, L. (2007). Father-daughter relationship as a moderator of sexual imprinting: A facialmetric study. *Evol Hum Behav, 28*(4), 248–52.

Can you tell if a man has daddy potential from his face alone?

Roney, J. R., Hanson, K. N., Durante, K. M., & Maestripieri, D. (2006). Reading men's faces: Women's mate attractiveness judgments track men's testosterone and interest in infants. *Proc Biol Sci, 273*(1598), 2169–75.

How might your mom or dad's age influence your attraction to older faces?

Perrett, D. I., Penton-Voak, I. S., Little, A. C., Tiddeman, B. P., Burt, D. M., Schmidt, N., et al. (2002). Facial attractiveness judgments reflect learning of parental age characteristics. *Proc Biol Sci, 269*(1494), 873–80.

Why do blue-eyed men prefer blue-eyed women?

Anderson, K. G. (2006). How well does paternity confidence match actual paternity? Evidence from worldwide nonpaternity rates. *Curr Anthropol, 47*(3), 513–20.

Laeng, B., Mathisen, R., & Johnsen, J.-A. (2007). Why do blue-eyed men prefer women with the same eye color? *Behav Ecol, 61*(3), 371–84.

CHAPTER 2 FOLLOWING YOUR NOSE

What are pheromones, and do they exist in humans?

Wysocki, C. J., & Preti, G. (2004). Facts, fallacies, fears, and frustrations with human pheromones. *Anat Rec A Discov Mol Cell Evol Biol, 281*(1), 1201–11.

Grammer, K., Fink, B., & Neave, N. (2005). Human pheromones and sexual attraction. *Eur J Obstet Gynecol Reprod Biol, 118*(2), 135–42.

Why do some men smell better to you than others?

Wedekind, C., Seebeck, T., Bettens, F., & Paepke, A. J. (1995). MHC-dependent mate preferences in humans. *Proc Biol Sci, 260*(1359), 245–49.

Ober, C., Weitkamp, L. R., Cox, N., Dytch, H., Kostyu, D., & Elias, S. (1997). HLA and mate choice in humans. *Am J Hum Genet, 61*(3), 497–504.

Rikowski, A., & Grammer, K. (1999). Human body odour, symmetry and attractiveness. *Proc Biol Sci, 266*(1422), 869–74.

Wedekind, C., & Penn, D. (2000). MHC genes, body odours, and odour preferences. *Nephrol Dial Transplant, 15*(9), 1269–71.

Havlíček, J., Roberts, S. C., & Flegr, J. (2005). Women's preference for dominant male odour: Effects of menstrual cycle and relationship status. *Biol Lett, 1*(3), 256–59.

Santos, P. S., Schinemann, J. A., Gabardo, J., & Bicalho Mda, G. (2005). New evidence that the MHC influences odor perception in humans: A study with 58 Southern Brazilian students. *Horm Behav, 47*(4), 384–88.

Milinski, M. (2006). The major histocompatibility complex, sexual selection, and mate choice. *Annu Rev Evol Syst, 37*, 159–86.

Why might you cheat on a man if you don't like his smell?

Ober, C., Hyslop, T., & Hauck, W. W. (1999). Inbreeding effects on fertility in humans: Evidence for reproductive compensation. *Am J Hum Genet, 64*(1), 225–31.

Grammer, K., Fink, B., & Neave, N. (2005). Human pheromones and sexual attraction. *Eur J Obstet Gynecol Reprod Biol, 118*(2), 135–42.

Garver-Apgar, C. E., Gangestad, S. W., Thornhill, R., Miller, R. D., & Olp, J. J. (2006). Major histocompatibility complex alleles, sexual responsivity, and unfaithfulness in romantic couples. *Psychol Sci, 17*(10), 830–35.

Milinski, M. (2006). The major histocompatibility complex, sexual selection, and mate choice. *Annu Rev Evol Syst, 37*, 159–86.

Why might the smell of men's sweat brighten your mood and senses?

Jacob, S., Kinnunen, L. H., Metz, J., Cooper, M., & McClintock, M. K. (2001). Sustained human chemosignal unconsciously alters brain function. *Neuroreport, 12*(11), 2391–94.

Lundstrom, J. N., Goncalves, M., Esteves, F., & Olsson, M. J. (2003). Psychological effects of subthreshold exposure to the putative human pheromone 4, 16-androstadien-3-one. *Horm Behav, 44*(5), 395–401.

Preti, G., Wysocki, C. J., Barnhart, K. T., Sondheimer, S. J., & Leyden, J. J. (2003). Male axillary extracts contain pheromones that affect pulsatile secretion of luteinizing hormone and mood in women recipients. *Biol Reprod, 68*(6), 2107–13.

Bensafi, M., Tsutsui, T., Khan, R., Levenson, R. W., & Sobel, N. (2004).

Sniffing a human sex-steroid derived compound affects mood and autonomic arousal in a dose-dependent manner. *Psychoneuroendocrinology, 29*(10), 1290–99.

Lundstrom, J. N., & Olsson, M. J. (2005). Subthreshold amounts of social odorant affect mood, but not behavior, in heterosexual women when tested by a male, but not a female, experimenter. *Biol Psychol, 70*(3), 197–204.

Jacob, T. J., Wang, L., Jaffer, S., & McPhee, S. (2006). Changes in the odor quality of androstadienone during exposure-induced sensitization. *Chem Senses, 31*(1), 3–8.

Keller, A., Zhuang, H., Chi, Q., Vosshall, L. B., & Matsunami, H. (2007). Genetic variation in a human odorant receptor alters odour perception. *Nature, 449*(7161), 468–72.

Wyart, C., Webster, W. W., Chen, J. H., Wilson, S. R., McClary, A., Khan, R. M., et al. (2007). Smelling a single component of male sweat alters levels of cortisol in women. *J Neurosci, 27*(6), 1261–65.

Does women's body odor have any effect on men?

Platek, S. M., Burch, R. L., & Gallup, G. B. (2001). Sex differences in olfactory self-recognition. *Physiol Behav, 73*(4), 535–40.

Singh, D., & Bronstad, P. M. (2001). Female body odour is a potential cue to ovulation. *Proc Biol Sci, 268*(1469), 797–801.

Kuukasjarvi, S., Eriksson, C. J., et al. (2004). Attractiveness of women's body odors over menstrual cycle: The role of oral contraceptives and receiver sex. *Behav Ecol, 15*(4), 579–84.

Havlíček, J., Dvorakova, R., et al. (2006). Non-advertised does not mean concealed: body odour changes across the human menstrual cycle. *Ethology, 112*, 81–90.

Can you tell people's sexual orientation by their smell?

Martins, Y., Preti, G., Crabtree, C. R., Runyan, T., Vainius, A. A., & Wysocki, C., J. (2005). Preference for human body odors is influenced by gender and sexual orientation. *Psychol Sci, 16*(9), 694–701.

Savic, I., Berglund, H., & Lindstrom, P. (2005). Brain response to putative pheromones in homosexual men. *Proc Natl Acad Sci U S A, 102*(20), 7356–61.

Berglund, H., Lindstrom, P., & Savic, I. (2006). Brain response to putative pheromones in lesbian women. *Proc Natl Acad Sci U S A, 103*(21), 8269–74.

Why might your sex drive pick up around breast-feeding women?

McClintock, M. K. (1971). Menstrual synchrony and suppression. *Nature, 229*(5282), 244–45.

Jacob, S., Spencer, N. A., Bullivant, S. B., Sellergren, S. A., Mennella, J. A., & McClintock, M. K. (2004). Effects of breastfeeding chemosignals on the human menstrual cycle. *Hum Reprod, 19*(2), 422–29.

Spencer, N. A., McClintock, M. K., Sellergren, S. A., Bullivant, S., Jacob, S., & Mennella, J. A. (2004). Social chemosignals from breastfeeding women increase sexual motivation. *Horm Behav, 46*(3), 362–70.

What might your perfume reveal about you?

Milinski, M., & Wedekind, C. (2001). Evidence for MHC-correlated perfume preferences in humans. *Behav Ecol, 12*(2), 140–49.

Milinski, M. (2006). The major histocompatibility complex, sexual selection, and mate choice. *Annu Rev Evol Syst, 37*, 159–86.

Can the smell of food or a fragrance be a turn-on?

Hirsch, A., & Gruss, J. (1999). Human male sexual response to olfactory stimuli. *J Neurol Orthop Med Surg 19*(1), 514–19.

Graham, C. A., Janssen, E., & Sanders, S. A. (2000). Effects of fragrance on female sexual arousal and mood across the menstrual cycle. *Psychophysiology, 37*(1), 76–84.

Herz, R., Beland, S., et al. (2004). Changing odor hedonic perception through emotional associations in humans. *Int J Comp Psych, 17*(4), 315–38.

Is a meat diet a turn-off?

Havlíček, J., & Lenochova, P. (2006). The effect of meat consumption on body odor attractiveness. *Chem Senses, 31*(8), 747–52.

Why might you seem less good-looking in a bad-smelling place?

Demattè, M. L., Osterbauer, R., & Spence, C. (2007). Olfactory cues modulate facial attractiveness. *Chem Senses, 32*(6), 603–10.

CHAPTER 3 A SOUND CHOICE

Why might deep-voiced men have more babies?

Thornhill, R., Gangestad, S. W., & Comer, R. (1995). Human female orgasm and mate fluctuating asymmetry. *Anim Behav, 50*(6), 1601–15.

Collins, S. A. (2000). Men's voices and women's choices. *Anim Behav, 60*(6), 773–80.

Hughes, S., Dispenza, F., et al. (2004). Ratings of voice attractiveness predict sexual behavior and body configuration. *Evol Hum Behav, 25*, 295–304.

Bruckert, L., Lienard, J. S., Lacroix, A., Kreutzer, M., & Leboucher, G. (2006). Women use voice parameters to assess men's characteristics. *Proc Biol Sci, 273*(1582), 83–89.

Evans, S., Neave, N., & Wakelin, D. (2006). Relationships between vocal characteristics and body size and shape in human males: An evolutionary explanation for a deep male voice. *Biol Psychol, 72*(2), 160–63.

Apicella, C. L., Feinberg, D. R., & Marlowe, F. W. (2007). Voice pitch predicts reproductive success in male hunter-gatherers. *Biol Lett., 3*(6), 682–84.

Puts, D. A., Hodges, C. R., et al. (2007). Men's voices as dominance signals: Vocal fundamental and formant frequencies influence dominance attributions among men. *Evol Hum Behav, 28*, 340–44.

Feinberg, D. R., DeBruine, L. M., Jones, B. C., & Little, A. C. (2008). Correlated preferences for men's facial and vocal masculinity. *Evol Hum Behav, 29*(4), 233–41.

Why do men prefer high-pitched female voices?

Hughes, S., Dispenza, F., et al. (2004). Ratings of voice attractiveness predict sexual behavior and body configuration. *Evol Hum Behav, 25,* 295–304.

Feinberg, D. R., Jones, B. C., et al. (2005). The voice and face of a woman: One ornament that signals quality? *Evol Hum Behav, 26,* 398–408.

Feinberg, D. R., Jones, B. C., Law Smith, M. J., Moore, F. R., DeBruine, L. M., Cornwell, R. E., et al. (2006). Menstrual cycle, trait estrogen level, and masculinity preferences in the human voice. *Horm Behav, 49*(2), 215–22.

Jones, B. C., Feinberg, D. R., DeBruine, L. M., Little, A. C., & Vukovic, J. (2008). Integrating cues of social interest and voice pitch in men's preferences for women's voices. *Biol Lett, 4*(2), 192–94.

How does your verbal "body language" reveal your attraction to someone?

Pentland, A. S. (2005). Socially aware computation and communication. *Computer,* March 63–69.

Can you tell if a person is gay by the sound of his or her voice?

Jacobs, G., Smyth, R., & Rogers, H. (2001). Language and sexuality: Searching for phonetic correlates of gay- and straight-sounding male voices. *Toronto Working Papers in Linguistics, 18,* 46–64.

Pierrehumbert, J. B., Bent, T., Munson, B., Bradlow, A. R., & Bailey, J. M. (2004). The influence of sexual orientation on vowel production. *J Acoust Soc Am, 116*(4 Pt 1), 1905–08.

Does the sound of your name affect how others perceive your looks?

Erwin, P. G. (1993). First names and perceptions of physical attractiveness. *J Psychol, 127*(6), 625–31.

Hopkin, M. (2004). Name game increases sex appeal. *News@Nature* Aug 10. http://www.npg.nature.com. Retrieved July 10, 2007.

Perfors, A. (2004). What's in a name? The effect of sounds symbolism on perception of facial attractiveness. Paper presented at CogSci 2004, Aug 5–7; Chicago, Ill.

CHAPTER 4 THE RACY PARTS

Why is long hair sexy?

Muscarella, F., & Cunningham, M. (1996). The evolutionary significance and social perception of male pattern baldness and facial hair. *Ethol Sociobiol, 17*(2), 99–117.

Hinsz, V. B., Matz, D. C., & Patience, R. A. (2001). Does women's hair signal reproductive potential? *J Exp Soc Psychol, 37*, 166–72.

Mesko, N., & Bereczkei, T. (2004). Hairstyle as an adaptive means of displaying phenotypic quality. *Hum Nat, 15*(3), 251–70.

Do gentlemen really prefer blondes?

Thelen, T. H. (1983). Minority type human mate preference. *Soc Biol, 30*(2), 162–80.

Rich, M. K., & Cash, T. F. (1993). The American image of beauty: media representations of hair color for four decades. *Sex Roles, 29*, 113–24.

Frost, P. (2006). European hair and eye color: A case of frequency-dependent sexual selection? *Evol Hum Behav, 27*, 85–103.

Hitsch, G. J., Hortaçsu, A., & Ariely, D. (2006). What makes you click? Mate preferences and matching outcomes in online dating. MIT Sloan Research Paper no. 4603–06.

Sorokowski, P. (2006). Do men prefer blondes? The influence of hair colour on the perception of age and attractiveness of women. *Studia Psychologiczne, 44*(3), 77–88.

Bry, C., Follenfant, A., & Meyer, T. (2007). Blonde like me: When self-construals moderate stereotype priming effects on intellectual performance. *J Exp Soc Psychol, 44*(3), 751–57.

Do tall men have prettier girlfriends?

Pawlowski, B., Dunbar, R. I., & Lipowicz, A. (2000). Tall men have more reproductive success. *Nature, 403*(6766), 156.

Nettle, D. (2002). Height and reproductive success in a cohort of British men. *Hum Nature, 13*, 473–91.

Nettle, D. (2002). Women's height, reproductive success, and the evolution of sexual dimorphism in modern humans. *Proc Biol Sci, 269*, 1919–23.

Pawlowski, B., & Jasieńska, G. (2005). Women's preferences for sexual dimorphism in height depend on menstrual cycle phase and expected duration of relationship. *Biol Psychol, 70*(1), 38–43.

Hitsch, G. J., Hortaçsu, Ali, & Ariely, Dan (2006). What makes you click? Mate preferences and matching outcomes in online dating. MIT Sloan Research Paper no. 4603–06.

Why are high heels sexy?

Pokrywka, L., Cabric, M., et al. (2006). Body mass index and waist:hip ratio are not enough to characterize female attractiveness. *Perception, 35*(12), 1693–97.

Swami, V., Einon, D., et al. (2006). The leg-to-body ratio as a human aesthetic criterion. *Body Image, 3*(4), 317–23.

Sorokowski, P., & Pawlowski, B. (2008). Adaptive preferences for leg length in a potential partner. *Evol Hum Behav, 29*(2), 86–91.

What does a "wiggle" in your walk reveal?

Johnson, K. L., & Tassinary, L. G. (2007). Compatibility of basic social perceptions determines perceived attractiveness. *Proc Natl Acad Sci U S A, 104*(12), 5246–51.

Provost, M. P., Quinsey, V. L., & Troje, N. F. (2007). Differences in gait across the menstrual cycle and their attractiveness to men. *Arch Sex Behav.*

Provost, M. P., Troke, N., & Quinsey, V. (2008). Short-term mating strategies and attraction to masculinity in point-light walkers. *Evol Hum Behav, 29*, 65–69.

Why are curves sexy?

Singh, D. (1993). Adaptive significance of female physical attractiveness: Role of waist-to-hip ratio. *J Pers Soc Psychol, 65*(2), 293–307.

Singh, D. (1994). Ideal female body shape: Role of body weight and waist-to-hip ratio. *Int J Eat Disord, 16*(3), 283–88.

Singh, D. (1994). Waist-to-hip ratio and judgment of attractiveness and healthiness of female figures by male and female physicians. *Int J Obes Relat Metab Disord, 18*(11), 731–37.

Singh, D. (2002). Female mate value at a glance: Relationship of waist-to-hip ratio to health, fecundity and attractiveness. *Neuro Endocrinol Lett, 23 Suppl 4*, 81–91.

Jasieńska, G., Ziomkiewicz, A., Ellison, P. T., Lipson, S. F., & Thune, I. (2004). Large breasts and narrow waists indicate high reproductive potential in women. *Proc Biol Sci, 271*(1545), 1213–17.

Lassek, W. D., & Gaulin, S. J. (2008). Waist-hip ratio and cognitive ability: Is gluteofemoral fat a privileged store of neurodevelopmental resources? *Evol Hum Behav, 29*(2), 26–34.

Why do men love big breasts?

Morris, D. (1967). *The Naked Ape: A Zoologist's Study of the Human Animal*. New York: Bantam Books.

Cant, J. (1981). Hypothesis for the evolution of human breasts and buttocks. *The American Naturalist, 117*(2), 199–204.

Møller, A. P., Soler, M., & Thornhill, R. (1995). Breast asymmetry, sexual selection and human reproductive success. *Ethol Sociobiol, 16*, 207–19.

Jasieńska, G., Ziomkiewicz, A., Ellison, P. T., Lipson, S. F., & Thune, I. (2004). Large breasts and narrow waists indicate high reproductive potential in women. *Proc Biol Sci, 271*(1545), 1213–17.

Scutt, D., Lancaster, G. A., & Manning, J. T. (2006). Breast asymmetry and predisposition to breast cancer. *Breast Cancer Res, 8*(2), R14.

Why do women feel pressured to be superskinny?

Singh, D. (1993). Adaptive significance of female physical attractiveness: Role of waist-to-hip ratio. *J Pers Soc Psychol, 65*(2), 293–307.

Singh, D. (1994). Ideal female body shape: Role of body weight and waist-to-hip ratio. *Int J Eat Disord, 16*(3), 283–88.

Singh, D. (1994). Waist-to-hip ratio and judgment of attractiveness and healthiness of female figures by male and female physicians. *Int J Obes Relat Metab Disord, 18*(11), 731–37.

Singh, D. (1994). Body fat distribution and perception of desirable female body shape by young black men and women. *Int J Eat Disord, 16*(3), 289–94.

Singh, D. (1995). Female judgment of male attractiveness and desirability for relationships: Role of waist-to-hip ratio and financial status. *J Pers Soc Psychol, 69*(6), 1089–1101.

Etcoff, N. (1999). *Survival of the Prettiest*. New York: Doubleday.

Tovée, M. J., Maisey, D. S., Emery, J. L., & Cornelissen, P. L. (1999). Visual cues to female physical attractiveness. *Proc Biol Sci, 266*(1415), 211–18.

Pope, H. G., Jr., Gruber, A., et al. (2000). Body image perception among men in three countries. *Am J Psych, 157*(8), 1297–1301.

Singh, D. (2002). Female mate value at a glance: Relationship of waist-to-hip ratio to health, fecundity and attractiveness. *Neuro Endocrinol Lett, 23 Suppl 4*, 81–91.

Frederick, D. A., Fessler, D. M., & Haselton, M. G. (2005). Do representations of male muscularity differ in men's and women's magazines? *Body Image, 2*(1), 81–86.

Swami, V., & Tovée, M. J. (2005). Female physical attractiveness in Britain and Malaysia. *Body Image, 2*, 115–28.

Swami, V., & Tovée, M. J. (2006). Does hunger influence judgments of female physical attractiveness? *Br J Psychol, 97*(Pt 3), 353–63.

Swami, V., & Tovée, M. J. (2007). Perceptions of female body weight and shape among indigenous and urban Europeans. *Scand J Psychol, 48*(1), 43–50.

Why do men feel pressured to be buff?

Frederick, D. A., Fessler, D. M., & Haselton, M. G. (2005). Do representations of male muscularity differ in men's and women's magazines? *Body Image, 2*(1), 81–86.

Yang, C. F., Gray, P., & Pope, H. G., Jr. (2005). Male body image in Taiwan versus the West: Yanggang Zhiqi meets the Adonis complex. *Am J Psychiatry, 162*(2), 263–69.

Frederick, D. A., & Haselton, M. G. (2007). Why is muscularity sexy?: tests of the fitness indicator hypothesis. *Pers Soc Psychol/Bull 33*(8), 1167–83.

Why do hungry men prefer heavier women?

Pettijohn, T. F., II, & Jungeberg, B. J. (2004). *Playboy Playmate* curves: changes in facial and body feature preferences across social and economic conditions. *Pers Soc Psychol/Bull, 30*(9), 1186–97.

Nelson, L. D., & Morrison, E. L. (2005). The symptoms of resource scarcity: judgments of food and finances influence preferences for potential partners. *Psychol Sci, 16*(2), 167–73.

Swami, V., & Tovée, M. J. (2006). Does hunger influence judgments of female physical attractiveness? *Br J Psychol, 97*(Pt 3), 353–63.

Little, A. C., Cohen, D. L., & Jones, B. C. (2007). Human preferences for facial masculinity change with relationship type and environmental harshness. *Behav Evol Sociobiol, 61*, 967–73.

Swami, V., Greven, C., & Furnham, A. (2007). More than just skin-deep? A pilot study integrating physical and non-physical factors in the perception of physical attractiveness. *Pers Individ Dif, 42*, 563–72.

Why do so many men wish they had bigger penises?

Ponchietti, R., Mondaini, N., Bonafe, M., Di Loro, F., Biscioni, S., & Masieri, L. (2001). Penile length and circumference: A study on 3,300 young Italian males. *Eur Urol, 39*(2), 183–86.

Francken, A. B., van de Wiel, H. B., van Driel, M. F., & Weijmar Schultz, W. C. (2002). What importance do women attribute to the size of the penis? *Eur Urol, 42*(5), 426–31.

Lever, J., Frederick, D. A., et al. (2006). Does size matter? Men's and women's views on penis size across the lifespan. *Psychol Men & Mascul, 7*(3), 129–43.

Are men with large genitals more likely to cheat?

Simmons, L. W., Firman, R. C., et al. (2004). Human sperm competition: Testis size, sperm production, and rates of extrapair competition. *Anim Behav, 68*, 297–302.

What's the purpose of pubic hair?

Harris Interactive (2006). Vagisil Women's Health Center Survey. Feb 2–6.

AskMen.com. Great Males Survey (2005). Retrieved April 20, 2008, from http://www.askmen.com/media_kit/survey/survey_index.html.

CHAPTER 5 HIS-AND-HERS HORMONES

Why are there days when men seem especially drawn to you?

Bullivant, S. B., Sellergren, S. A., Stern, K., Spencer, N. A., Jacob, S., Mennella, J. A., et al. (2004). Women's sexual experience during the menstrual cycle: Identification of the sexual phase by noninvasive measurement of luteinizing hormone. *J Sex Res, 41*(1), 82–93.

Roberts, S. C., Havlíček, J., Flegr, J., Hruskova, M., Little, A. C., Jones, B. C., et al. (2004). Female facial attractiveness increases during the fertile phase of the menstrual cycle. *Proc Biol Sci, 271 Suppl 5*, S270–72.

Jones, B. C., Perrett, D. I., Little, A. C., Boothroyd, L., Cornwell, R. E., Feinberg, D. R., et al. (2005). Menstrual cycle, pregnancy and oral contraceptive use alter attraction to apparent health in faces. *Proc Biol Sci, 272*(1561), 347–54.

Gizewski, E. R., Krause, E., Karama, S., Baars, A., Senf, W., & Forsting, M. (2006). There are differences in cerebral activation between females in distinct menstrual phases during viewing of erotic stimuli: A fMRI study. *Exp Brain Res, 174*(1), 101–8.

Smith, M. J., Perrett, D. I., Jones, B. C., Cornwell, R. E., Moore, F. R.,

Feinberg, D. R., et al. (2006). Facial appearance is a cue to oestrogen levels in women. *Proc Biol Sci, 273*(1583), 135–40.

Haselton, M. G. (2007). Male sexual attractiveness predicts differential ovulatory shifts in female extra-pair attraction and male mate retention. *Evol Hum Behav, 27*, 247–58.

Miller, G. F., Tybur, J. M., & Jordan, B. D. (2007). Ovulatory cycle effects on tip earnings by lap dancers: Economic evidence for human estrus? *Evol Hum Behav 28* (6), 375–81.

Why are you sometimes drawn to macho guys even if they're not your type?

Little, A. C., Jones, B. C., Penton-Voak, I. S., Burt, D. M., & Perrett, D. I. (2002). Partnership status and the temporal context of relationships influence human female preferences for sexual dimorphism in male face shape. *Proc Biol Sci, 269*(1496), 1095–1100.

Gangestad, S. W., Simpson, J. A., Cousins, A. J., Garver-Apgar, C. E., & Christensen, P. N. (2004). Women's preferences for male behavioral displays change across the menstrual cycle. *Psychol Sci, 15*(3), 203–7.

Havlíček, J., Roberts, S. C., & Flegr, J. (2005). Women's preference for dominant male odour: Effects of menstrual cycle and relationship status. *Biol Lett, 1*(3), 256–59.

Cornwell, R. E., Law Smith, M. J., Boothroyd, L. G., Moore, F. R., Davis, H. P., Stirrat, M., et al. (2006). Reproductive strategy, sexual development and attraction to facial characteristics. *Philos Trans R Soc Lond B Biol Sci, 361*(1476), 2143–54.

Pillsworth, E. G., & Haselton, M. G. (2006). Male sexual attractiveness predicts differential ovulatory shifts in extra-pair attraction and male mate retention. *Evol Hum Behav, 27*, 247–58.

Gangestad, S. W., Garver-Apgar, C. E., Simpson, J. A., & Cousins, A. J. (2007). Changes in women's mate preferences across the ovulatory cycle. *J Pers Soc Psychol, 92*(1), 151–63.

Little, A. C., & Perrett, D. I. (2007). Using composite images to assess accuracy in personality attribution to faces. *Br J Psychol, 98*(Pt 1), 111–26.

Roney, J. R., & Simmons, Z. L. (2008). Women's estradiol predicts preference for facial cues of men's testosterone. *Horm Behav, 53*(1), 14–19.

Why don't people go into heat like other animals?

Hrdy, S. B. (1977). Infanticide as a primate reproductive strategy. *Am Sci, 65*(1), 40–49.

Alexander, R. D., & Noonan, K. (1979). Concealment of ovulation, parental care, and human social evolution. In N. A. Chagnon & W. Irons (Eds.), *Evolutionary Biology and Human Social Behavior*, 436–53. North Scituate, MA: Duxbury.

Hrdy, S. B. (1990). Sex bias in nature and in history. *Yearbook of Physical Anthropology, 33*, 25–37.

Diamond, J. (1997). *Why Is Sex Fun? The Evolution of Human Sexuality*. New York: Basic Books.

Sillén-Tullberg, B., & Møller, A. (1993). The relationship between concealed ovulation and mating systems in anthropoid primates. *Am Nat, 141*, 1–25.

How do the seasons affect your sex life?

Eriksson, A. W., & Fellman, J. (2000). Seasonal variation of livebirths, stillbirths, extramarital births and twin maternities in Switzerland. *Twin Res, 3*(4), 189–201.

Andersson, A. M., Carlsen, E., Petersen, J. H., & Skakkebaek, N. E. (2003). Variation in levels of serum inhibin B, testosterone, estradiol, luteinizing hormone, follicle-stimulating hormone, and sex hormone-binding globulin in monthly samples from healthy men during a 17-month period: Possible effects of seasons. *J Clin Endocrinol Metab, 88*(2), 932–37.

James, M. (2005). Summertime mystery: More born, fewer die in August. Retrieved 3 March, 2008, from http://abcnews.go.com/Health/Science/story?id=945911.

van Anders, S. M., Hampson, E., & Watson, N. V. (2006). Seasonality, waist-to-hip ratio, and salivary testosterone. *Psychoneuroendocrinology, 31*(7), 895–99.

Macdowall W., & Wellings, K. (2008). Summer nights: A review of the evidence of seasonal variations in sexual health indications among young people. *Health Ed, 108*(1), 40–53.

What can you tell about people by the ratio of their fingers?

Manning, J. T., Scutt, D., Wilson, J., & Lewis-Jones, D. I. (1998). The ratio of 2nd to 4th digit length: A predictor of sperm numbers and concentrations of testosterone, luteinizing hormone and oestrogen. *Hum Reprod, 13*(11), 3000–3004.

Manning, J. T., Barley, L., Walton, J., Lewis-Jones, D. I., Trivers, R. L., Singh, D., et al. (2000). The 2nd: 4th digit ratio, sexual dimorphism, population differences, and reproductive success: Evidence for sexually antagonistic genes? *Evol Hum Behav, 21*(3), 163–83.

Manning, J. T. (2002). The ratio of 2nd to 4th digit length and performance in skiing. *J Sports Med Phys Fitness, 42*(4), 446–50.

Fink, B., Neave, N., & Manning, J. T. (2003). Second to fourth digit ratio, body mass index, waist-to-hip ratio, and waist-to-chest ratio: Their relationships in heterosexual men and women. *Ann Hum Biol, 30*(6), 728–38.

Fink, B., Grammer, K., Mitteroecker, P., Gunz, P., Schaefer, K., Bookstein, F. L., et al. (2005). Second to fourth digit ratio and face shape. *Proc Biol Sci, 272*(1576), 1995–2001.

Manning, J. T., Fink, B., Neave, N., & Caswell, N. (2005). Photocopies yield lower digit ratios (2D:4D) than direct finger measurements. *Arch Sex Behav, 34*(3), 329–33.

Honekopp, J., Manning, J. T., & Muller, C. (2006). Digit ratio (2D:4D) and physical fitness in males and females: Evidence for effects of prenatal androgens on sexually selected traits. *Horm Behav, 49*(4), 545–49.

Honekopp, J., Voracek, M., & Manning, J. T. (2006). 2nd to 4th digit ratio

(2D:4D) and number of sex partners: Evidence for effects of prenatal testosterone in men. *Psychoneuroendocrinology, 31*(1), 30–37.

Kraemer, B., Noll, T., Delsignore, A., Milos, G., Schnyder, U., & Hepp, U. (2006). Finger length ratio (2D:4D) and dimensions of sexual orientation. *Neuropsychobiology, 53*(4), 210–14.

Manning, J. T., Morris, L., & Caswell, N. (2007). Endurance running and digit ratio (2D:4D): Implications for fetal testosterone effects on running speed and vascular health. *Am J Hum Biol, 19*(3), 416–21.

Peters, M., Manning, J. T., & Reimers, S. (2007). The effects of sex, sexual orientation, and digit ratio (2D:4D) on mental rotation performance. *Arch Sex Behav, 36*(2), 251–60.

Why do men lose their judgment and decision-making skills when looking at pretty women?

Wilson, M., & Daly, M. (2004). Do pretty women inspire men to discount the future? *Proc Biol Sci, 271 Suppl 4*, S177–79.

Ariely, D., & Lowenstein, G. (2006). The heat of the moment: The effect of sexual arousal on sexual decision making. *J Behav Decis Making, 19*(2), 87–98.

Van den Bergh, B., & Dewitte, S. (2006). Digit ratio (2D:4D) moderates the impact of sexual cues on men's decisions in ultimatum games. *Proc Biol Sci, 273*(1597), 2091–95.

Why does doing something dangerous or exciting increase attraction?

Dutton, D. G., & Aron, A. P. (1974). Some evidence for heightened sexual attraction under conditions of high anxiety. *J Pers Soc Psychol, 30*(4), 510–17.

Meston, C. M., & Frohlich, P. F. (2003). Love at first fright: Partner salience moderates roller-coaster-induced excitation transfer. *Arch Sex Behav, 32*(6), 537–44.

Why do you like and trust a guy more after you've been intimate (even just cuddling)?

Young, L. J., & Wang, Z. (2004). The neurobiology of pair bonding. *Nat Neurosci, 7*(10), 1048–54.

Grewen, K. M., Girdler, S. S., Amico, J., & Light, K. C. (2005). Effects of partner support on resting oxytocin, cortisol, norepinephrine, and blood pressure before and after warm partner contact. *Psychosom Med, 67*(4), 531–38.

Kosfeld, M., Heinrichs, M., Zak, P. J., Fischbacher, U., & Fehr, E. (2005). Oxytocin increases trust in humans. *Nature, 435*(7042), 673–76.

Fisher, H. E., Aron, A., & Brown, L. L. (2006). Romantic love: A mammalian brain system for mate choice. *Philos Trans R Soc Lond B Biol Sci, 361*(1476), 2173–86.

Why do men mellow out when they're in a relationship?

Mazur, A. (1995). Biosocial models of deviant behaviour among male army veterans. *Biol Psychol, 41*(3), 271–93.

Mazur, A., & Booth, A. (1998). Testosterone and dominance in men. *Behav Brain Sci, 21*(3), 353–63; discussion 363–97.

Storey, A. E., Walsh, C. J., Quinton, R. L., & Wynne-Edwards, K. E. (2000). Hormonal correlates of paternal responsiveness in new and expectant fathers. *Evol Hum Behav, 21*(2), 79–95.

Burnham, T. C., Chapman, J. F., Gray, P. B., McIntyre, M. H., Lipson, S. F., & Ellison, P. T. (2003). Men in committed, romantic relationships have lower testosterone. *Horm Behav, 44*(2), 119–22.

Gray, P. B. (2003). Marriage, parenting, and testosterone variation among Kenyan Swahili men. *Am J Phys Anthropol, 122*(3), 279–86.

Gray, P. B., Campbell, B. C., Marlowe, F. W., Lipson, S. F., & Ellison, P. T. (2004). Social variables predict between-subject but not day-to-day variation in the testosterone of US men. *Psychoneuroendocrinology, 29*(9), 1153–62.

Gray, P. B., Yang, C. F., & Pope, H. G., Jr. (2006). Fathers have lower salivary testosterone levels than unmarried men and married non-fathers in Beijing, China. *Proc Biol Sci, 273*(1584), 333–39.

McIntyre, M., Gangestad, S. W., Gray, P. B., Chapman, J. F., Burnham, T. C., O'Rourke, M. T., et al. (2006). Romantic involvement often reduces men's testosterone levels—but not always: The moderating role of extrapair sexual interest. *J Pers Soc Psychol, 91*(4), 642–51.

When are you most committed to your partner?

Fisher, M. L. (2004). Female intrasexual competition decreases female facial attractiveness. *Proc Biol Sci, 271 Suppl 5*, S283–85.

DeBruine, L. M., Jones, B. C., et al. (2005). Women's attractiveness judgments of self-resembling faces change across the menstrual cycle. *Horm Behav, 47*(4), 379–83.

Jones, B. C., Little, A. C., Boothroyd, L., DeBruine, L. M., Feinberg, D. R., Smith, M. J., et al. (2005). Commitment to relationships and preferences for femininity and apparent health in faces are strongest on days of the menstrual cycle when progesterone level is high. *Horm Behav, 48*(3), 283–90.

CHAPTER 6 SIGNS AND SIGNALS

What body language do women use to express interest?

Moore, M. M. (1985). Nonverbal courtship patterns in women: Rejection signaling. *Semiotica, 118* (3/4), 201–14.

Moore, M. M. (1985). Nonverbal courtship patterns in women: Context and consequences. *Ethol Sociobiol, 6*, 236–46.

What's the strongest signal you can use to get someone's attention?

Strack, F., Martin, L. L., & Stepper, S. (1988). Inhibiting and facilitating conditions of the human smile: A nonobtrusive test of the facial feedback hypothesis. *J Pers Soc Psychol, 54*(5), 768–77.

Dabbs, J. M., Jr. (1997). Testosterone, smiling, and facial appearance. *J Nonverbal Behav, 21*(1), 45–55.

Adams, R. B., Jr., & Kleck, R. E. (2003). Perceived gaze direction and the processing of facial displays of emotion. *Psychol Sci, 14*(6), 644–47.

Jones, B. C., DeBruine, L. M., Little, A. C. (2006). Integrating gaze direction and expression in preference for attractive faces. *Psychol Sci, 17* (7), 588–91.

Winston, J. S., O'Doherty, J., Kilner, J. M., Perrett, D. I., & Dolan, R. J. (2007). Brain systems for assessing facial attractiveness. *Neuropsychologia, 45*(1), 195–206.

Conway, C. A., Jones, B. C., DeBruine, L. M. (2008). Evidence for adaptive design in human gaze preference. *Proc Biol Sci, 275*(1630), 63–69.

What exactly makes a smile attractive?

Strack, F., Martin, L. L., & Stepper, S. (1988). Inhibiting and facilitating conditions of the human smile: A nonobtrusive test of the facial feedback hypothesis. *J Pers Soc Psychol, 54*(5), 768–77.

Frank, M. G., Ekman, P., & Friesen, W. V. (1993). Behavioral markers and recognizability of the smile of enjoyment, *J Pers Soc Psychol, 64*(1), 83–93.

Dabbs, J. M., Jr. (1997). Testosterone, smiling, and facial appearance. *J Nonverbal Behav, 21*(1), 45–55.

Why do guys think you're into them when you're just being friendly?

Abbey, A. (1987). Misperceptions of friendly behavior as sexual interest. *J Pers Soc Psychol, 42*, 830–38.

Haselton, M. G., & Buss, D. M. (2000). Error management theory: A new perspective on biases in cross-sex mind reading. *J Pers Soc Psychol, 78*(1), 81–91.

Haselton, M. G. (2003). The sexual overperception bias: Evidence of a systematic bias in men from a survey of naturally occurring events. *J Res Pers, 37*, 34–47.

Henningsen, D. (2004). Flirting with meaning: An examination of miscommunication in flirting interactions. *Sex Roles, 50*(7/8), 481–99.

Hamann, S., Herman, R. A., Nolan, C. L., & Wallen, K. (2004). Men and women differ in amygdala response to visual sexual stimuli. *Nat Neurosci*, *7*(4), 411–16.

Levesque, M. (2006). Toward an understanding of gender differences in inferring sexual interest. *Psychol Women Quart*, *30*, 150–58.

What body language do guys use to get your attention?

Renninger, L., Wade, J., & Grammer, K. (2004). Getting that female glance: Patterns and consequences of male nonverbal behavior in courtship consequences. *Evol Human Behav* *25*(6), 416–31.

How persuasive is a touch?

Chapell, M. S., Beltran, W., Santanello, M., Takahashi, M., Bantom, S. R., Donovan, J. S., et al. (1999). Men and women holding hands: II. Whose hand is uppermost? *Percept Mot Skills*, *89*(2), 537–49.

Guéguen, N. (2007). Courtship compliance: The effect of touch on women's behavior. *Soc Influence*, *2*, 81–97.

What's the hidden agenda in men's pickup lines?

Bale, C., & Morrison, R. (2006). Chat-up lines as male sexual displays. *Pers Indiv Dif*, *40*, 655–64.

Cooper, M., O'Donnell, D. O., et al. (2007). Chat-up lines as male displays: Effects of content, sex, and personality. *Pers Indiv Dif*, *43*, 1075–85.

Why is blushing sexy?

Shields, S., Mallory, M. E., et al. (1990). The experience and symptoms of blushing as a function of age and reported frequency of blushing. *J Nonverbal Behav*, *14*(3), 171–87.

Changizi, M. A., Zhang, Q., & Shimojo, S. (2006). Bare skin, blood and the evolution of primate colour vision. *Biol Lett*, *2*(2), 217–21.

Why does mimicry make you more likable?

Chartrand, T. L., & Bargh, J. A. (1999). The chameleon effect: The perception-behavior link and social interaction. *J Pers Soc Psychol*, *76*(6), 893–910.

Neumann, R., & Strack, F. (2000). "Mood contagion": The automatic transfer of mood between persons. *J Pers Soc Psychol, 79*(2), 211–23.

Cheng, C. M., & Chartrand, T. L. (2003). Self-monitoring without awareness: Using mimicry as a nonconscious affiliation strategy. *J Pers Soc Psychol, 85*(6), 1170–79.

Lakin, J. L., & Chartrand, T. L. (2003). Using nonconscious behavioral mimicry to create affiliation and rapport. *Psychol Sci, 14*(4), 334–39.

van Baaren, R. B., Holland, R. W., Kawakami, K., & van Knippenberg, A. (2004). Mimicry and prosocial behavior. *Psychol Sci, 15*(1), 71–74.

Lee, T. W., Josephs, O., Dolan, R. J., & Critchley, H. D. (2006). Imitating expressions: Emotion-specific neural substrates in facial mimicry. *Soc Cogn Affect Neurosci, 1*(2), 122–35.

Why do you turn your head to the right when you kiss?

Nicholls, M. E., Clode, D., Wood, S. J., & Wood, A. G. (1999). Laterality of expression in portraiture: Putting your best cheek forward. *Proc Biol Sci, 266*(1428), 1517–22.

Güntürkün, O. (2003). Human behaviour: Adult persistence of head-turning asymmetry. *Nature, 421*(6924), 711.

Nicholls, M. E., Ellis, B. E., Clement, J. G., & Yoshino, M. (2004). Detecting hemifacial asymmetries in emotional expression with three-dimensional computerized image analysis. *Proc Biol Sci, 271*(1540), 663–68.

Barrett, D., Greenwood, J. G., & McCullagh, J. F. (2006). Kissing laterality and handedness. *Laterality, 11*(6), 573–79.

Why do we French-kiss?

Singh, D., & Bronstad, P. M. (2001). Female body odour is a potential cue to ovulation. *Proc Biol Sci, 268*(1469), 797–801.

Kimata, H. (2006). Kissing selectively decreases allergen-specific IgE production in atopic patients. *J Psychosom Res, 60*(5), 545–47.

Hughes, S. M., Harrison, M.A., & Gallup, G. G., Jr. (2007). Sex differences in romantic kissing among college students. *Evol Psychol, 5*(3), 612–31.

CHAPTER 7 SEX AND SEDUCTION

Why do men have more casual sex?

Clarke, R., & Hatfield, E. (1989). Gender differences in receptivity to sexual offers. *J Psychol Human Sex*, 2, 39–55.

Buss, D. M., & Schmitt, D. P. (1993). Sexual strategies theory: An evolutionary perspective on human mating. *Psychol Rev*, *100*(2), 204–32.

Schmitt, D. P., Alcalay, L., Allik, J., Ault, L., Austers, I., Bennett, K. L., et al. (2003). Universal sex differences in the desire for sexual variety: Tests from 52 nations, 6 continents, and 13 islands. *J Pers Soc Psychol*, *85*(1), 85–104.

Li, N. P., & Kenrick, D. T. (2006). Sex similarities and differences in preferences for short-term mates: What, whether, and why. *J Pers Soc Psychol*, *90*(3), 468–89.

Why are fewer men than women bisexual?

Chivers, M. L., Rieger, G., Latty, E., & Bailey, J. M. (2004). A sex difference in the specificity of sexual arousal. *Psychol Sci*, *15*(11), 736–44.

Rieger, G., Chivers, M. L., & Bailey, J. M. (2005). Sexual arousal patterns of bisexual men. *Psychol Sci*, *16*(8), 579–84.

Lippa, R. A. (2006). Is high sex drive associated with increased sexual attraction to both sexes? It depends on whether you are male or female. *Psychol Sci*, *17*(1), 46–52.

Lippa, R. A. (2007). The relation between sex drive and sexual attraction to men and women: A cross-national study of heterosexual, bisexual, and homosexual men and women. *Arch Sex Behav*, *36*(2), 209–22.

Are men more aroused than women by pornography?

Wilson, G. D. (1987). Male-female differences in sexual activity, enjoyment, and fantasies. *J Pers Ind Dif*, *8*(1), 125–27.

Holstege, G., Georgiadis, J. R., Paans, A. M., Meiners, L. C., van der Graaf, F. H., & Reinders, A. A. (2003). Brain activation during human male ejaculation. *J Neurosci*, *23*(27), 9185–93.

Canli, T., & Gabrieli, J. D. (2004). Imaging gender differences in sexual arousal. *Nat Neurosci*, *7*(4), 325–26.

Chivers, M. L., Rieger, G., Latty, E., & Bailey, J. M. (2004). A sex difference in the specificity of sexual arousal. *Psychol Sci, 15*(11), 736–44.

Hamann, S., Herman, R. A., Nolan, C. L., & Wallen, K. (2004). Men and women differ in amygdala response to visual sexual stimuli. *Nat Neurosci, 7*(4), 411–16.

Khamai, R. (2006). Women become aroused as quickly as men. *New Scientist*; 2 October.

Can a romantic movie set the mood for love?

Schultheiss, O. C., Wirth, M. M., & Stanton, S. J. (2004). Effects of affiliation and power motivation arousal on salivary progesterone and testosterone. *Horm Behav*, 46(5), 592–99.

Are good dancers also good in bed?

Thornhill, R., & Gangestad, S. W. (1994). Human fluctuating asymmetry and sexual behavior. *Psychol Sci, 5*(3), 297–302.

Thornhill, R., Gangestad, S. W., & Comer, R. (1995). Human female orgasm and mate fluctuating asymmetry. *Anim Behav, 50*(6), 1601–1615.

Brown, W. M., Cronk, L., Grochow, K., Jacobson, A., Liu, C. K., Popovic, Z., et al. (2005). Dance reveals symmetry especially in young men. *Nature, 438*(7071), 1148–50.

Is chocolate really an aphrodisiac?

Salonia, A., Fabbri, F., Zanni, G., Scavini, M., Fantini, G. V., Briganti, A., et al. (2006). Chocolate and women's sexual health: An intriguing correlation. *J Sex Med, 3*(3), 476–82.

How does alcohol affect your sex life?

Steele, C. M., & Josephs, R. A. (1990). Alcohol myopia: Its prized and dangerous effects. *Am Psychol, 45*(8), 921–33.

Lindman, R. E., Koskelainen, B., & Eriksson, C. J. P. (1999). Drinking, menstrual cycle, and female sexuality: A diary study. *Alcoholism 23*(1), 169–73.

Can semen make you happier?

Gallup, G. G., Jr., Burch, R. L., & Platek, S. M. (2002). Does semen have antidepressant properties? *Arch Sex Behav, 31*(3), 289–93.

Burch, R. L., & Gallup, G. (2006). *The Psychobiology of Human Semen.* New York: Cambridge University Press.

Why do women have orgasms?

Thornhill, R., Gangestad, S. W., & Comer, R. (1995). Human female orgasm and mate fluctuating asymmetry. *Anim Behav, 50*(6), 1601–15.

Singh, D., Meyer, W., Zambarano, R. J., & Hurlbert, D. F. (1998). Frequency and timing of coital orgasm in women desirous of becoming pregnant. *Arch Sex Behav, 27*(1), 15–29.

Baker, R., & Bellis, M. (1993). Human sperm competition: Ejaculate adjustment by males and the function of masturbation. *Anim Behav, 46*(5), 861–85.

Are orgasms genetic?

Dawood, K., Kirk, K. M., Bailey, J. M., Andrews, P. W., & Martin, N. G. (2005). Genetic and environmental influences on the frequency of orgasm in women. *Twin Res Hum Genet, 8*(1), 27–33.

Do women really reach their sexual peak in their thirties?

Baumeister, R. F. (2000). Gender differences in erotic plasticity: The female sex drive as socially flexible and responsive. *Psychol Bull, 126*(3), 347–74; discussion 347–89.

Schmitt, D. P., Shackelford, T. K., & Buss, D. (2002). Is there an early-30s peak in female sexual desire? Cross-sectional evidence from the United States and Canada. *Can J Hum Sex, 11*(1), 1–18.

Why is intercourse more satisfying than masturbation?

Komisaruk, B. R., & Whipple, B. (1998). Love as sensory stimulation: Physiological consequences of its deprivation and expression. *Psychoneuroendocrinology, 23*(8), 927–44.

Komisaruk, B. R., Whipple, B., Crawford, A., Liu, W. C., Kalnin, A., & Mosier, K. (2004). Brain activation during vaginocervical self-stimulation and orgasm in women with complete spinal cord injury: fMRI evidence of mediation by the vagus nerves. *Brain Res, 1024*(1–2), 77–88.

Brody, S. (2006). Blood pressure reactivity to stress is better for people who recently had penile-vaginal intercourse than for people who had other or no sexual activity. *Biol Psychol, 71*(2), 214–22.

Brody, S., & Kruger, T. H. (2006). The post-orgasmic prolactin increase following intercourse is greater than following masturbation and suggests greater satiety. *Biol Psychol, 71*(3), 312–15.

Do men and women experience orgasm the same way?

Vance, E. B., & Wagner, N. N. (1976). Written descriptions of orgasm: A study of sex differences. *Arch Sex Behav, 5*(1), 87–98.

Bolen, J. G. (1980). The male orgasm: Pelvic contractions measured by anal probe. *Arch Sex Behav, 9*(6), 508–21.

Mah, K., & Binik, Y. M. (2001). The nature of human orgasm: A critical review of major trends. *Clin Psychol Rev, 21*(6), 823–56.

Komisaruk, B. R., Whipple, B., Crawford, A., Liu, W. C., Kalnin, A., & Mosier, K. (2004). Brain activation during vaginocervical self-stimulation and orgasm in women with complete spinal cord injury: fMRI evidence of mediation by the vagus nerves. *Brain Res, 1024*(1–2), 77–88.

Holstege, G., Georgiadis, J. R., Paans, A. M., Meiners, L. C., van der Graaf, F. H., & Reinders, A. A. (2003). Brain activation during human male ejaculation. *J Neurosci, 23*(27), 9185–93.

Bianchi-Demicheli, F., & Ortigue, S. (2007). Toward an understanding of the cerebral substrates of woman's orgasm. *Neuropsychologia, 45*(12), 2645–59.

Ortigue, S., Grafton, S. T., & Bianchi-Demicheli, F. (2007). Correlation between insula activation and self-reported quality of orgasm in women. *Neuroimage, 37*(2), 551–60.

Why do people with satisfying sex lives still masturbate?

Janus, S., & Janus, C. (1993). *The Janus Report on Sexual Behavior*. New York: Wiley.

Baker, R., & Bellis, M. (1993). Human sperm competition: Ejaculate adjustment by males and the function of masturbation. *Anim Behav 46*(5), 861–85.

Das, A. (2007). Masturbation in the United States. *J Sex Marital Ther, 33*(4), 301–17.

Why aren't you sexually attracted to people who grew up with you?

Shepher, J. (1971). Mate selection among second generation kibbutz adolescents and adults: Incest avoidance and negative imprinting. *Arch Sex Behav, 1*, 293–307.

Wolf, A. P. (1995). *Sexual Attraction and Childhood Association: A Chinese Brief for Edward Westermarck*. Stanford, CA: Stanford Univ Press.

Lieberman, D., Tooby, J., & Cosmides, L. (2003). Does morality have a biological basis? An empirical test of the factors governing moral sentiments relating to incest. *Proc Biol Sci, 270*(1517), 819–26.

Lieberman, D., Tooby, J., & Cosmides, L. (2007). The architecture of human kin detection. *Nature, 445*(7129), 727–31.

CHAPTER 8 THE DATING MIND-SET

What do women and men value in a partner?

Buss, D. M. (1989). Sex differences in human mate preferences: Evolutionary hypotheses tested in 37 cultures. *Behav Brain Sci, 12*, 1–49.

Buss, D. M. (2000). Desires in human mating. *Ann N Y Acad Sci, 907*, 39–49.

Buss, D. M., & Shackelford, T. K. (2008). Attractive women want it all: Good genes, economic investment, parenting proclivities, and emotional commitment. *Evol Psychol, 6*(1), 134–46.

McNulty, J. K., Neff, L. A., & Karney, B. R. (2008). Beyond initial attraction: Physical attractiveness in newlywed marriage. *J Fam Psych, 22*(1); 135–43.

Pawlowski, B., & Jasieńska, G. (2008). Women's body morphology and preference for sexual partners' characteristics. *Evol Hum Behav*, *29*(1) 19–25.

What secret biases do data from online dating sites reveal?

Hitsch, G. J., Hortaçsu, A., & Ariely, D. (2006). What makes you click? Mate preferences and matching outcomes in online dating. MIT Sloan Research Paper no. 4603–6.

Fisman, R., Iyengar, S., Kamenica, E., & Simonson, I. (2006). Racial preference in dating: Evidence from a speed dating experience. Working paper; Columbia Business School.

Hancock, J., Toma, C., et al. (2007). The truth about lying in online dating profiles. *Computer/Human Interaction*; May.

Why might single men spend more and single women volunteer more?

Roney, J. R. (2003). Effects of visual exposure to the opposite sex: Cognitive aspects of mate attraction in human males. *Pers Soc Psychol Bull*, *29*(3), 393–404.

Griskevicius, V., Tybur, J. M., Sundie, J. M., Cialdini, R. B., Miller, G. F., & Kenrick, D. T. (2007). Blatant benevolence and conspicuous consumption: When romantic motives elicit strategic costly signals. *J Pers Soc Psychol*, *93*(1), 85–102.

Haselton, M. G., Buss, D. M., Oubaid, V., & Angleitner, A. (2005). Sex, lies, and strategic interference: The psychology of deception between the sexes. *Pers Soc Psychol Bull*, *31*(1), 3–23.

Why do men give women fancy dinners and vacations instead of useful gifts?

Sozou, P. D., & Seymour, R. M. (2005). Costly but worthless gifts facilitate courtship. *Proc Biol Sci*, *272*(1575), 1877–84.

Why does creativity get men laid?

Buss, D. M., & Barnes, M. (1986). Preferences in human mate selection. *J Pers Soc Psychol, 50*(3), 559–80.

Miller, G. (2001). *The Mating Mind: How Sexual Choice Shaped the Evolution of Human Nature*. New York: Anchor.

Li, N. P., Bailey, J. M., Kenrick, D. T. (2002). The necessities and luxuries of mate preferences. *J Pers Soc Psychol, 82,* 947–55.

Nettle, D., & Clegg, H. (2006). Schizotypy, creativity, and mating success in humans. *Proc Biol Sci, 97*(P 2), 177–90.

Anderson, K. G. (2006). How well does paternity confidence match actual paternity? Evidence from worldwide nonpaternity rates. *Curr Anthropol, 47*(3), 513–20.

Haselton, M. G. (2007). Male sexual attractiveness predicts differential ovulatory shifts in female extra-pair attraction and male mate retention. *Evol Hum Behav, 27,* 247–58.

Why aren't there more male muses?

Buss, D. M., & Barnes, M. (1986). Preferences in human mate selection. *J Pers Soc Psychol, 50*(3), 559–80.

Miller, G. (2001). *The Mating Mind: How Sexual Choice Shaped the Evolution of Human Nature*. New York: Anchor.

Griskevicius, V., Tybur, J. M., Sundie, J. M., Cialdini, R. B., Miller, G. F., & Kenrick, D. T. (2007). Blatant benevolence and conspicuous consumption: When romantic motives elicit strategic costly signals. *J Pers Soc Psychol, 93*(1), 85–102.

Why is humor a turn-on?

Fraley, B., & Aron, A. (2004). The effect of a shared humorous experience on closeness in initial encounters. *Pers Relationship, 11,* 61–78.

Azim, E., Mobbs, D., Jo, B., Menon, V., & Reiss, A. L. (2005). Sex differences in brain activation elicited by humor. *Proc Natl Acad Sci U S A, 102*(45), 16496–501.

Bazzini, D., Stack, E., et al. (2007). The effect of reminiscing about laughter on relationship satisfaction. *Motiv Emotion*, 31(1), 25–34.

Why are you more attracted to picky people (and they to you)?

Eastwick, P. W., Finkel, E. J., et al. (2007). Selective versus unselective romantic desire: not all reciprocity is created equal. *Psychol Sci 18*(4), 317–19.

Jones, B. C., DeBruine, L. M., Little, A. C., Burriss, R. P., & Feinberg, D. R. (2007). Social transmission of face preferences among humans. *Proc Biol Sci, 274*(1611), 899–903.

Little, A. C., Burriss, R., et al. (2008). Social influence in human face preference: Men and women are influenced more for long-term than short-term attractiveness decisions. *Evol Hum Behav, 29*(2), 140–46.

Why do people seem hotter when others are into them?

Brown, G. R., & Fawcett, T. W. (2005). Sexual selection: Copycat mating in birds. *Curr Biol, 15*(16), R626–28.

Swaddle, J. P., Cathey, M. G., Correll, M., & Hodkinson, B. P. (2005). Socially transmitted mate preferences in a monogamous bird: A non-genetic mechanism of sexual selection. *Proc Biol Sci, 272*(1567), 1053–58.

Jones, B. C., DeBruine, L. M., Little, A. C., Burriss, R. P., & Feinberg, D. R. (2007). Social transmission of face preferences among humans. *Proc Biol Sci, 274*(1611), 899–903.

Little, A. C., Burriss, R., et al. (2008). Social influence in human face preference: Men and women are influenced more for long-term than short-term attractiveness decisions. *Evol Hum Behav, 29*(2), 140–46.

Does a guy love you less after looking at (other) beautiful women?

Kenrick, D. T., & Gutierres, S. E. (1980). Contrast effects and judgments of physical attractiveness. *J Pers Soc Psychol, 38*, 131–41.

Kenrick, D. T., Gutierres, S., & Goldberg, L. (1989). Influence of popular erotica on judgments of strangers and mates. *J Exp Soc Psychol, 25*, 159–67.

Kenrick, D. T., Neuberg, S. L., Zierk, K. L., & Krones, J. M. (1994). Evolution and social cognition: Contrast effects as a function of sex, dominance, and physical attractiveness. *Pers Soc Psychol Bull, 20*, 210–17.

Kanazawa, S., & Still, M. C. (2000). Teaching may be hazardous to your marriage. *Evol Hum Behav, 21*(3), 185–90.

Mishra, S., Clark, A., & Daly, M. (2007). One woman's behavior affects the attractiveness of others. *Evol Hum Behav, 28*, 145–49.

Why shouldn't you spill everything about yourself on a first date?

Gibbs, J., Ellison, N., & Heino, R. (2006). Self-presentation in online personals. *Commun Res, 33*(2), 152–77.

Norton, M. I., Frost, J. H., & Ariely, D. (2007). Less is more: The lure of ambiguity, or why familiarity breeds contempt. *J Pers Soc Psychol, 92*(1), 97–105.

Why do you overestimate your competition?

Hill, S. E. (2007). Overestimation bias in mate competition. *Evol Hum Behav, 28*, 118–23.

Is there an ideal number of people to date before you settle down?

Brooks, M. (2000). What's love got to do with it? *New Scientist*. Oct 28.

Simão, J., & Todd, P. M. (2003). Emergent patterns of mate choice in human populations. *Artif Life, 9*(4), 403–17.

Todd, P. M. (2007). Coevolved cognitive mechanisms in mate search: Making decisions in a decision-shaped world. In J. P. Forgas, M. G. Haselton, & W. von Hippel (Eds.), *Evolution and the Social Mind: Evolutionary Psychology and Social Cognition*. New York: Psychology Press.

CHAPTER 9 LOVE ON THE BRAIN

How does being passionately in love change your brain?

Fisher, H., Aron, A., Mashek, D., Li, H., Strong, G., & Brown, L. L. (2002). The neural mechanisms of mate choice: A hypothesis. *Neuro Endocrinol Lett, 23 Suppl 4*, 92–97.

Marazziti, D., & Canale, D. (2004). Hormonal changes when falling in love. *Psychoneuroendocrinology, 29*(7), 931–36.

Aron, A., Fisher, H., Mashek, D. J., Strong, G., Li, H., & Brown, L. L. (2005). Reward, motivation, and emotion systems associated with early-stage intense romantic love. *J Neurophysiol, 94*, 327–37.

Fisher, H., Aron, A., & Brown, L. L. (2005). Romantic love: An fMRI study of a neural mechanism for mate choice. *J Comp Neurol, 493*(1), 58–62.

Fisher, H. E., Aron, A., & Brown, L. L. (2006). Romantic love: A mammalian brain system for mate choice. *Philos Trans R Soc Lond B Biol Sci, 361*(1476), 2173–86.

Fisher, H., & Thomson, J. A., Jr. (2006). "Lust, romance, attachment: Do the side effects of serotonin-enhancing antidepressants jeopardize romantic love, marriage, and fertility?" In Platek, S., Keenan, J., & Shackelford, T. (Eds.), *Evolutionary Cognitive Neuroscience*. Cambridge, MA: MIT Press.

Why do you find your partner so amazing, even if no one else does?

Murray, S. L., Holmes, J. G., & Griffin, D. W. (1996). The self-fulfilling nature of positive illusions in romantic relationships: Love is not blind, but prescient. *J Pers Soc Psychol, 71*(6), 1155–80.

Gagne, F. M., & Lydon, J. E. (2004). Bias and accuracy in close relationships: An integrative review. *Pers Soc Psychol Rev, 8*(4), 322–38.

Geher, G., & Bloodworth, R. (2005). Motivational underpinnings of romantic partner perceptions. *J Soc Pers Relat, 22*(2), 255–81.

Why do couples look increasingly alike over time?

Zajonc, R. B., Adelmann, P. K., et al. (1987). Convergence in the physical appearance of spouses. *Motiv Emotion, 11*(4), 335–46.

Does being in love make you blind to other people's love?

Aloni, M., & Bernieri, F. (2004). Is love blind? The effects of experience and infatuation in the perception of love. *J Nonverbal Behav, 28*(4), 287–310.

Are people naturally monogamous?

Fisher, H. E., Aron, A., Mashek, D., Li, H., & Brown, L. L. (2002). Defining the brain systems of lust, romantic attraction, and attachment. *Arch Sex Behav, 31*(5), 413–19.

Dupanloup, I., Pereira, L., Bertorelle, G., Calafell, F., Prata, M. J., Amorim, A., et al. (2003). A recent shift from polygyny to monogamy in humans is suggested by the analysis of worldwide Y-chromosome diversity. *J Mol Evol, 57*(1), 85–97.

Why does absence make men's hearts grow fonder?

Thornhill, R., Gangestad, S. W., & Comer, R. (1995). Human female orgasm and mate fluctuating asymmetry. *Anim Behav, 50*(6), 1601–15.

Singh, D., Meyer, W., Zambarano, R. J., & Hurlbert, D. F. (1998). Frequency and timing of coital orgasm in women desirous of becoming pregnant. *Arch Sex Behav, 27*(1), 15–29.

Shackelford, T. K. (2002). Psychological adaptation to human sperm competition. *Evol Hum Behav, 23*, 123–38.

Baker, R., & Bellis, M. (1993). Human sperm competition: Ejaculate adjustment by males and the function of masturbation. *Anim Behav, 46*(5), 861–85.

Shackelford, T. K., Goetz, A. T., McKibbin, W. F., & Starratt, V. G. (2007). Absence makes the adaptations grow fonder: Proportion of time apart from partner, male sexual psychology, and sperm competition in humans (*Homo sapiens*). *J Comp Psychol, 121*(2), 214–20.

Buss, D. M., & Shackelford, T. K. (1997). From vigilance to violence: Mate retention tactics in married couples. *J Pers Soc Psychol, 72*(2), 346–61.

Buss, D. M. (2002). Human mate guarding. *Neuro Endocrinol Lett, 23 Suppl 4*, 23–29.

Pietrzak, R., Laird, J. D., Stevens, D. A., & Thompson, N. S. (2002). Sex differences in human jealousy. *Evol Hum Behav, 23*, 83–94.

Pillsworth, E. G., Haselton, M. G., & Buss, D. M. (2004). Ovulatory shifts in female sexual desire. *J Sex Res, 41*(1), 55–65.

Kilgallon, S. J., & Simmons, L. (2005). Image content influences men's semen quality. *Biol Lett, 1*(3), 253–55.

Pillsworth, E. G., & Haselton, M. G. (2006). Male sexual attractiveness predicts differential ovulatory shifts in extra-pair attraction and male mate retention. *Evol Hum Behav, 27*, 247–58.

Gallup, G. G., & Burch, R. L. (2006). Semen displacement as a sperm competition strategy. *Hum Nat, 17*(3), 253–64.

Do your genes make you more faithful or adulterous?

Fisher, H., Aron, A., Mashek, D., Li, H., Strong, G., & Brown, L. L. (2002). The neural mechanisms of mate choice: A hypothesis. *Neuro Endocrinol Lett, 23 Suppl 4*, 92–97.

Lim, M. M., Murphy, A. Z., & Young, L. J. (2004). Ventral striatopallidal oxytocin and vasopressin V1a receptors in the monogamous prairie vole (*Microtus ochrogaster*). *J Comp Neurol, 468*(4), 555–70.

Lim, M. M., Wang, Z., Olazabal, D. E., Ren, X., Terwilliger, E. F., & Young, L. J. (2004). Enhanced partner preference in a promiscuous species by manipulating the expression of a single gene. *Nature, 429*(6993), 754–57.

Young, L. J., & Wang, Z. (2004). The neurobiology of pair bonding. *Nat Neurosci, 7*(10), 1048–54.

Fisher, H. E., Aron, A., & Brown, L. L. (2006). Romantic love: A mammalian brain system for mate choice. *Philos Trans R Soc Lond B Biol Sci, 361*(1476), 2173–86.

Zeki, S. (2007). The neurobiology of love. *FEBS Lett, 581*(14), 2575–79.

How does your brain "grow" when you're in love?

Ortigue, S., Bianchi-Demicheli, F., Hamilton, A. F., & Grafton, S. T. (2007). The neural basis of love as a subliminal prime: An event-related functional magnetic resonance imaging study. *J Cogn Neurosci, 19*(7), 1218–30.

ACKNOWLEDGMENTS

I am very grateful to all the researchers whose experiments inspired this book. In particular, I thank those who took the time to answer nagging questions, send reprints, and/or read relevant sections of the manuscript: Stephanie Ortigue, Peter Frost, Gordon Gallup Jr., Christine Garver-Apgar, Meghan Provost, Suma Jacob, Meredith Chivers, Charles Wysocki, Randy Thornhill, Stuart Brody, Manfred Milinski, Sari van Anders, Sandy Pentland, Jorge Simão, Jan Havlíček, Debra Lieberman, Peter Gray, Ben Jones, Nicolas Guéguen, Claus Wedekind, Lee Cronk, Galit Yovel, Bruno Laeng, Nancy Kanwisher, Arthur Aron, and Monica Moore. Of course, any errors or omissions are my own.

I would particularly like to thank my editor, Danielle Perez, not only for her immediate attraction to this book idea but also for all her excellent edits and suggestions along the way. I also thank Belina Huey, Sue Warga, Barb Burg, Chris Artis, Catherine Leonardo, Bonnie Ammer, and Fabrizio LaRocca. Special thanks and appreciation go to my agent, Rick Broadhead.

With gratitude I acknowledge the kind permission from Katherine Larson and Kelly Clayton to excerpt their respective poems. Many thanks to Zenaide for her emotional support and enthusiasm for this book, and to other friends who have shared their dating stories, and to my parents for their encouragement (and use of their Vermont home).

Most of all, I thank my husband, Peter, who tells me he prefers brunettes and always raises my dopamine.

INDEX

ABOUT THE AUTHOR

Jena Pincott has a background in biology and was a production assistant on science documentaries for PBS. She is a former senior editor at Random House, and is the author of *Success: Advice for Achieving Your Goals from Remarkably Accomplished People* and *Healing: Advice for Recovering Your Strength and Spirit from the World's Most Famous Survivors.* She lives in New York City.